THE NUCLEAR-POWER REBELLION

ALSO BY RICHARD S. LEWIS

A Continent for Science
The Antarctic Adventure

Appointment on the Moon
The Full Story of Americans in Space, from Explorer 1 to the Lunar Landing and Beyond

The Nuclear-Power Rebellion | *CITIZENS VS. THE ATOMIC INDUSTRIAL ESTABLISHMENT*

RICHARD S. LEWIS

The Viking Press | *New York*

For Louise, Jonathan, and David,
and any that come after

Copyright © 1972 by Richard S. Lewis
All rights reserved
First published in 1972 by The Viking Press, Inc.
625 Madison Avenue, New York, N.Y. 10022
Published simultaneously in Canada by
The Macmillan Company of Canada Limited
SBN *670-51823-9*
Library of Congress catalog card number: 72-79005
Printed in U.S.A. by Vail-Ballou Press, Inc.

Contents

1 Another Kind of Fire 3
2 A Pattern for Plunder 26
3 The Radiation Heresy 48
4 The Ghost of Galileo 81
5 The Intervenors 109
6 Boot Hill 148
7 Plowshare 172
8 Bonanza from Bombs 199
9 The Battle of the Breeder 232
10 Milestone at Calvert Cliffs 259

Reference Notes 299

Index 307

THE NUCLEAR-POWER REBELLION

1
Another Kind of Fire

> The Committee has noted with concern in the past year the increased public opposition, not a little of it wholly unreasoning, to the construction of all electrical generating sources and their transmission systems.
> —*Report of the Joint Committee on Atomic Energy, United States Congress, 1970*

From the altitude where airliners pass over the region, the Endless Mountains of northeastern Pennsylvania are configured like the folds of a carpet pushed inward from its edge. The rounded ridges and shallow valleys run in parallel from the southwestern to the northeastern horizons. The view from aloft supports a theory that the mountains were formed by forces exerted against the continental margins of North America by sea-floor spreading, as magma from the earth's mantle was extruded from the mid-Atlantic Ocean rift.

Through this northerly region of Appalachia flows the north branch of the Susquehanna River. Along its east bank, 1300 acres of level farmland near the village of Meshoppen in Wyoming County were staked out in 1967 for a historic development. On this site, the Pennsylvania Electric Company, a subsidiary of General Public Utilities Corporation, proposed to build an advanced type of atomic-power plant in partnership with the Atomics International Division of North American Rockwell Corporation and the United States Atomic Energy Commission. The proposed plant would demonstrate for the first time the commercial application of the Liquid Metal Fast Breeder Reactor to electric-power production. A second-generation fission reactor which the AEC had developed at a cost approaching a half-billion dollars to supplant its light-water reactor, then coming into widespread use, the "breeder" was the product of a makeshift national energy policy based on the expectation that, early in the twenty-first century, advanced fission reactors would be producing half of the nation's electrical energy and eventually would replace coal, oil, and gas to become the principal energy source.

In fuel efficiency and economy, the Liquid Metal Fast Breeder Reactor (LMFBR) would be as far ahead of the first light-water reactor put into commercial demonstration in 1957 at Shippingport, Pennsylvania, as that device was ahead of Enrico Fermi's Chicago Pile No. 1, where the first controlled nuclear chain reaction was achieved on December 2, 1942.

A nuclear reactor is another kind of fire—a source of heat. Instead of coal, oil, or gas, the light-water reactor developed in the United States "burns" the fissionable fuel uranium-235. The heat of the fissioning atoms boils water and makes steam to run turbine generators. But electric utilities using these reactors face rising fuel costs because uranium-235 is very limited in nature. It amounts to only 0.7 per cent of natural uranium.

With breeder reactors, however, the picture is brighter. The fast breeder can tap up to 75 per cent of the energy in natural uranium—a hundred times more than the light-water reactors can utilize. Breeders do this by transmuting the bulk of natural uranium, which is not fissionable, into fissionable plutonium-239, which does not exist in nature. In theory, the breeder reactor can "breed" three atoms of plutonium-239 for every two atoms of uranium-235 undergoing fission in the fuel rods. Hence it is capable of breeding more fuel than it consumes.

With the breeder reactor, the nuclear-power industry could look forward to the extension of an economical fuel supply from decades to millennia. The breeder thus becomes the device which a growing nuclear-power establishment can employ to become a principal supplier of energy in the United States and the world.

A Rising of Skeptics

In proposing to build the first demonstration breeder in the Endless Mountains, the utility did not anticipate the reaction of some of the residents. The communities in Wyoming County resemble those of New England in some ways. White-painted frame houses, unpretentious and neat, are predominant. A considerable degree of economic homogeneity is suggested by the lack of conspicuous affluence or poverty. In this part of Pennsylvania, on the border of the great anthracite coal-mining region, as in New England, one encounters a highly articulate skepticism that has questioned the impact of every technological innovation since the electric light.

Possibly because the summers are pleasant and the autumns are long and warm, with mountains illuminated by glorious foliage of red and gold, the natives take great satisfaction in the natural surroundings that they have begun to refer to as their environment. They display a special concern for the preservation of a style of uncrowded country and village life in which the devices of modern technology can be used without altering or intruding coarsely on the rustic scene.

It was inevitable that some of the residents would consider the construction of a large nuclear-power demonstration plant, still in an experimental stage of design, as an intrusion and a threat in an area where no need for it existed. Five hundred megawatts was more than five times the demand for electricity in the region. Who needed that much? The answer was obvious and unacceptable to some people. Not Wyoming County or even the centers of nearby Scranton and Wilkes-Barre. This was power for the future, serving the Northeastern industrial complex. It had no relation to the needs of Wyoming County, and its environmental impact was considered only by the people who lived there.

It was by no means an overnight development. As early as 1962, the blueprint for the Meshoppen breeder and its sister demonstrators had been drawn in the AEC's report to President John F. Kennedy. It said, in part, ". . . we estimate that by A.D. 2000, nuclear power would be assuming the total increase in electrical-energy production. . . . We have crudely estimated that by the century's end, nuclear installations might actually be generating approximately half of the total electric energy in the country." 1*
The report, which in the words of Chairman Glenn T. Seaborg represented a "new and hard look at the role of nuclear power in our economy," urged the development of the breeder reactor to exploit "the vast energy resources latent" in uranium-238 and thorium.

Some of the people who began to feel concern about the plans of Pennsylvania Electric Company, Atomics International, and the AEC for Wyoming County had never heard of this report, were not aware of industry claims of greatly increased power requirements for the twenty-first century, and were not impressed by the AEC's visions of the atomic future. The demonstrator served interests remote in space and in time, interests which were guarded by mountain walls and distance from any accidental release of massive radiation from a new type of nuclear reactor. And it served these interests, it seemed, at the expense of the security of a small, dispersed population which the nuclear-energy planners considered evacuable in the "unlikely event" (as they were always saying) of a nuclear emergency.

What kind of an emergency would that be? Some people believed that nuclear reactors could blow up, like bombs. And they were right—about fast reactors, fueled with plutonium. Under certain conditions, breeders running on plutonium could blow up and spew deadly radioactive poisons all over the countryside. Or they could be sabotaged or struck by missiles with the same effect.

In a report on "Public Health Factors in Reactor Site Selection," the United States Public Health Service had commented:

---

* Numbered reference notes begin on p. 299.

Population density is an important factor to consider in selecting optimum reactor sites. Based on data available to the public, release rates of radioactive air and water contaminants from power reactors during normal operations have been quite low. However, in the unlikely event of a reactor accident, relatively large quantities of radioactive materials could be released to the environment. This necessitates locating reactors in areas with low population densities or close to populations of high mobility who could be moved in a short time period.[2]

While the population of Wyoming County could not be described as "highly mobile," the road network in the county would enable most people to flee a massive release of radiation from a reactor at Meshoppen, if they were warned in time. How the urban populations of Scranton and Wilkes-Barre within a radius of thirty miles would fare might be regarded with less optimism. But Wyoming County clearly met the Public Health Service criteria for population density. The county seat, Tunkhannock, has a population of 2100 and a traffic light.

A Committee Is Formed

One evening in the autumn of 1969, Mrs. Joan Daniels, a housewife and mother who worked part time in the county library at Tunkhannock. was driving home from a Quaker meeting at Wilkes-Barre, when she decided that something ought to be done about the Meshoppen breeder. She turned to her friend and remarked, "I'd like to do something about that."

Mrs. Daniels recalled (even after two years) that her friend had responded without any hesitation, "All right. I'll help."

"That's the frame of mind you get into at a Quaker meeting," Mrs. Daniels related. So began an organization called the Citizens Committee for Environmental Concern, which in 1969 opened its campaign to block the construction of the 500-megawatt LMFBR at Meshoppen.

Within twelve months, the committee had recruited members in Luzerne and Lackawanna counties, had entered into an extended dialogue with state legislators and with representatives in Congress, had brought a two-months series of seminars on atomic energy to the Tunkhannock High School auditorium, and had gathered six thousand names to a petition asking county, state, and federal officials to prevent the construction of the Meshoppen breeder. The committee was influential in motivating State Representative Franklin L. Kury of Sunbury to introduce a bill in the 1970 Pennsylvania legislature outlawing the building of breeder reactors in that state. The Kury bill did not survive—but it was a beginning.

By the spring of 1970, the proliferation of conventional, lightwater reactors in Pennsylvania—a coal-producing state—had aroused concern among environmental groups in Pittsburgh and Philadelphia. According to the State Director of Radiological Health, Thomas M. Gerusky, Pennsylvania had more nuclear reactors in operation, in construction, or in the planning stages than any other state. In view of the rising concern about the Meshoppen breeder emanating from Tunkhannock, and the publicity attending Representative Kury's bill, the State Senate appointed a Select Committee to "see what all the hullabaloo is about." Public hearings were held at the State Capitol in Harrisburg during the summer and fall of 1970.

The Hullabaloo

The anatomy of a hullabaloo is least likely to be perceived at legislative hearings where, too often, ideas are drowned in rhetoric. One frosty night in January 1971, I attended a meeting of the Citizens Committee for Environmental Concern at the United States Department of Agriculture regional office in Tunkhannock. It was there that the nature of the hullabaloo became apparent. The snow-draped mountains gleamed softly under the bright stars like sheeted forms, and the snow crunched crisply in the parking lot, where automobiles, trucks, and station wagons drew up to bear witness to the potential mobility of the population in the event of a nuclear catastrophe.

About a dozen men and women were seated around a polished table, talking informally and all at once. The aroma of percolating coffee filled the room. What could these "country people" do to frustrate the design of an Establishment in which the government was formally allied with the private electric-power industry and its industrial suppliers? When the Limited Test Ban Treaty was debated in the United States Senate in 1963, some scientists expressed regret that the mysteries of the nuclear age were so arcane that only an intellectual elite, an atomic priesthood, was capable of understanding and passing judgment on questions of nuclear policy. Advancing nuclear technology was disenfranchising the people.

However reasonable this opinion may have appeared in 1963, there was no basis for it in Wyoming County in 1970. After a year of study, lectures, seminars, and discussion, members of the Citizens Committee for Environmental Concern had acquired a firm grasp of the technical issues of nuclear power and its safety problems. Indeed, I found that the folk of the Endless Mountains had a more sophisticated understanding of the hazards of radiation contamination and fuel-core overheating than most of their elected representatives. They could see clearly enough the propaganda, half-truths, and inconsistencies in the promotional matter purveyed by both the AEC and the industry to persuade residents of the area that fission reactors were safe, clean, wholesome devices for power production.

Mrs. Daniels had become co-chairman of the committee with Dr. Bryan Lee, Jr., a veterinarian. Mrs. Daniels, with the efficiency of a librarian, had assembled a prodigious file of atomic-energy and environmental information. With the forbearance of her husband, Sidney, she set up a library of her own in the bedroom of their home.

Members of the Citizens Committee were people who vote carefully at elections, pay their taxes, perform useful work in the community, help each other in times of stress, offer their children as much higher education as the children will take, and are concerned about the future of their community, nation, and humanity.

So much was obvious to any neutral observer. Committee members included Bryan Lee, Jr.'s father, Bryan, Sr., a dentist and county commissioner; Spencer Burr, the former county treasurer; Virginia Sheret, a retired United States Army dietician; Francis Heisler, the mayor of nearby Factoryville; Davis R. Hobbs, an attorney; Victor Capucci, Jr., and William Ohme, businessmen of Mehoopany; Tom Shelburne and Gus DiStadio of UHF television station WNEP, which had devoted public-service programing to the reactor issue; and two well-informed and public-spirited housewives, Mrs. Angela Rinehimer and Mrs. Betty Tewksbury.

The Challenge

The discussion that evening at Tunkhannock revealed the basic conflict between a concerned citizenry and a technical establishment. The citizens questioned the right of an establishment to impose upon them a new technology affecting their health and safety without their consent. Here was the basis of public intervention into the plans of a powerful new force—the Atomic Industrial Establishment—to install an advanced atomic-power system, based on a plutonium-fuel cycle of great potential hazard, throughout the country. Wyoming County residents had been alerted early to this development. Some of them reacted by asserting the right to be heard of the demos—in which the collective wisdom of a democratic society must reside. It was a challenge which the Establishment's technical elite would evade or resist as long as it could.

An important issue plaguing the citizens, not only in Wyoming County, Pennsylvania, but also in Maryland and Michigan, Wisconsin and California, Minnesota and Colorado, was the credibility of the Establishment. "There is so much contradictory evidence about the effects of these reactors among scientists at the highest level that we don't know what to believe," one of the housewives said. "They rattle around in their academic armor and they arrive at diametrically opposed conclusions from the same data. Scientists are supposed to be honest. How can they be, when they are working for a company or a government agency and depending on it for a living?"

Bryan Lee, Jr., said: "Maybe they could operate the plant safely, but a whole series of such plants could be hazardous. I think the way we all feel is . . . we just don't want to live near one."

Gus DiStadio questioned the AEC's radiation standards:

> Do you think any human being has the right to set a radiation standard for me? To determine how much radiation I should get? I have to accept the natural radiation background that God gave me, but I don't accept man-made radiation. As a species, we have evolved with a certain radiation background. What happens to our species when more radiation is added? Can they tell us that?

Others spoke of their fears of sabotage against a nuclear-power plant. One had written to Defense Secretary Melvin R. Laird asking how vulnerable reactors are to sabotage or to a missile attack that would loose massive radiation from the reactor core over the countryside. The inquiry had received a reply: There were no measures to protect nuclear-power plants per se from a missile attack.

Discussion followed on that point. Should nuclear reactors, especially those in the East, within range of submarine-launched ballistic missiles, be built underground, like missile silos? Who in the government considered such contingencies? No one knew.

Several mothers said that "propaganda" had been distributed to school children purporting to "educate" the children, and through them their parents, about the advantages of atomic energy. One of the pamphlets brought home by the children had been prepared by the AEC's Division of Technical Information, to tell all about nuclear energy and the good things the friendly, workaday atom could do for everybody. The mothers said they feared the children were being indoctrinated to accept radiation hazards, or the AEC's view of them, as the natural order of things. One of the pamphlets prepared by an industrial firm described radioactive nuclear waste as "goop." Everybody knew what "goop" was—nothing to worry about.

In the initial stages of its campaign against the project, the Citizens Committee found their state and Congressional representatives attentive. But the span of attention turned out to be brief. As the committee persisted in demanding answers and action from their elected representatives, the responses of these officials became more formalized and terse. Members of the committee said that utility public-relations people who had initially catered to inquiries became increasingly indifferent and remote as the inquiries became more challenging.

"They regard us as a bunch of country people who are in over our heads," said Bryan Lee, Jr. "It happens to be true. We *are* a bunch of country people, but we think that after a year of study, after a year of listening to some of the most brilliant scientists in the country on the subject, we have learned something about nuclear technology. We've had some of the best nuclear scientists and engineers from Penn State over here to lecture. We know enough at least to ask questions and to know when we're not getting the right answers."

The Hearing in Harrisburg

The Citizens Committee for Environmental Concern had raised the question of the consent of the electorate to exposure to the risks of nuclear-power facilities at the State Senate Select Committee hearing on atomic-power plants in Pennsylvania during the summer and fall of 1970. The hearing was clearly a political response to the concerns of citizens' groups in Philadelphia and Pittsburgh, as well as in Wyoming County, and to the concerns of the state's massive coal industry.

In its formal statement to the Senate Select Committee, the Citizens Committee for Environmental Concern asserted:

> We cannot condone the logic which advocates siting a reactor of any type in a low-population area at a time when it is considered

far too great a risk if sited in a highly populated area. We further submit that the risk we would be expected to tolerate is an indignity upon the rights of every citizen of the United States.

Lee, Jr., raised the consent issue in his testimony as a witness before the Select Committee:

What if the people don't want . . . nuclear generators? What if we don't want fuel-reprocessing plants [which also emit radioactive waste] in our state? What if the people prefer to receive their electricity from hydroelectric facilities in Canada that would have no environmental impact on anyone and which are assets to the development of that country? What if the people prefer increased research into the production and transmission of electricity from harnessing the energy of the tides and the sun? What if fusion power seems much more acceptable? If funds should be allocated for earnest research into the control of sulfur dioxide or for the development of magnetohydrodynamic power? We submit that the power industry needs to be washed and hung on a line. It needs to be gotten out in the open for the public's inspection."

An Unfriendly Giant

The controversy over the Meshoppen breeder revealed to the Citizens Committee the existence of an Atomic Industrial Establishment which somehow had the power to invade their environment and, from their point of view, threaten their security without their consent. The Establishment had first appeared in Pennsylvania in 1957 when the Shippingport pressurized-water reactor was developed as a joint demonstration project of a commercial atomic-power plant by the AEC, the Westinghouse Electric Company, and the Duquesne Light Company. A second commercial plant, built by the General Electric Company for the Commonwealth Edison Company of Chicago at Morris, Illinois, became operable in 1959. The following year, the Yankee Atomic Electric Company's pressurized-water reactor, built by Westinghouse, had gone on line at Rowe, Massachusetts. The Consumers Power Company started up a General Electric boiling-water reactor in 1962 at Big Rock, Michigan. In the same year Consolidated Edison Company began operating its first pressurized-water reactor, manufactured by Babcock and Wilcox, at Indian Point near Buchanan, New York. In 1963 the Pacific Gas and Electric Company's General Electric boiling-water reactor went into service in Humboldt Bay, California.

The trend toward nuclear power accelerated in the second half of the 1960s. By 1971 there were 21 operating light-water nuclear reactors in commercial service in the United States; 56 more were in construction, and 37 were on order—a total of 114, with an electrical-energy potential of 92,135,800 kilowatts. When the new generation of fast-breeder reactors appeared on the public horizon in 1967, the nuclear industry had shown promise of becoming a giant—an unfriendly giant from the viewpoint of people concerned about its environmental impact. Privately owned investor utilities and their industrial suppliers formed the economic base of the new Establishment, with the AEC acting sometimes as its agent and at other times as its management. Eventually, the Congressional Joint Committee on Atomic Energy became its board of directors.

The threat of the new breeder reactor to public health came early into focus in Pennsylvania where it was explored in the Senate Select Committee investigation into nuclear-reactor safety at public hearings between August and October 1970. Two safety issues were discussed at these hearings: first, preventing the fuel core of the reactor from becoming so overheated that it would melt, destroy its container, and eject radioactive debris over the surrounding area—a problem particularly menacing in the breeder reactor, which used plutonium in its fuel assembly, the most hazardous, perhaps, of all radioactive materials; and second, the hazard of radioactivity released during routine operation of the plant or during an emergency that did not result in a meltdown.

The likelihood of the fuel core becoming hot enough to melt through its containment had been studied for years. Several estimates of the magnitude of the disaster resulting from a meltdown had emerged, all indicating that the radiation effect would be analogous to the detonation of a nuclear bomb near an inhabited area. How likely was such an accident?

Bryan Lee, Jr., recalled for the State Senate Select Committee the meltdown of the Enrico Fermi experimental fast-breeder reactor, a forerunner of the larger one which the Establishment wanted to construct and demonstrate at Meshoppen. Fermi I, as the reactor was called, was a joint project of the AEC and the Detroit Edison Company. It was built in 1963 at Lagoona Beach, Michigan, and operated at low power for several years. In 1966, the liquid-sodium coolant was blocked and the core overheated and partially melted. There was no significant release of radioactivity, however, according to the AEC.

"It took almost a year to find out what triggered the accident, but eventually it was discovered," Lee, Jr., testified.

> By inserting a specially designed periscope into the interior of the reactor, a piece of crumpled metal about eight inches long was discovered. This piece of metal was part of one of three zirconium plates, which had been fastened to the cone at the bottom of the reactor to guide the sodium flow. These plates were not part of the original design and had been hurriedly added at the last moment, ironically as an extra safety measure. Because it was a last-minute, hurry-up job, it was never shown in the plans or work drawings submitted, so that no one in the plant at the time of the accident even knew of their existence. A former employee who had retired but was still employed as a consultant finally remembered the incident and was able to enlighten everyone. What happened was due to vibration and pressure. One of the plates had torn loose and blocked the flow of liquid-sodium coolant. Without the coolant, part of the fuel melted down. A great concern is that when this happens in a fast reactor, the fuel can then recongeal, become a critical mass, and set off an explosion which could rupture the core of the reactor and release large amounts of radioactivity to the air. Fortunately, this did not happen in this case, perhaps because the reactor was running at only a fraction of its potential power at the time of the accident.

Dresden II Incident

A lesser-known accident, which was not made public at the time it happened, occurred June 5, 1970, in the boiling-water reactor of Commonwealth Edison Company's Dresden II plant, southwest of Chicago. The reactor was being tested at 75 per cent of its power when a spurious signal in the pressure-control system altered the steam flow to the turbine. A series of malfunctions then followed and resulted in a massive discharge of steam and water into the reactor dry well. Radioactive iodine, which can cause cancer of the thyroid, became concentrated in the well to a hundred times the maximum permissible level. Radioactivity in the gas coming out of the smokestack increased from 10,000 to 25,000 microcuries per second for about thirty minutes and then subsided to the lower level. Although temperatures and pressures jumped in the reactor core, plant operators were able to control them, and the reactor was shut down until August 8, 1970.

The AEC reported a year later that no *significant* amount of radioactivity had been released to the environment.[3] Subsequently, I met with engineering and public-relations personnel of Commonwealth Edison and asked why no public announcement of the incident had been made earlier. The answer was that because the unit was in a testing phase, no one thought it was necessary. How many other instances which illustrate that accidents can and do happen in the "older" generation of nuclear plants have been withheld from the public?

The danger in this kind of incident is a rupture in the cooling system allowing the water that circulates around and cools the bundles of uranium-dioxide fuel elements to drain out. Without cooling water, normal operating temperature of about 315 degrees centigrade zooms to 1800 degrees in less than a minute. This is the melting point of the zirconium alloy in which the uranium is encased.

A backup emergency core-cooling system for light-water reactors to prevent a meltdown and the dispersal of massive radioactivity outside the plant has been in development for several years. At this writing, none has proved acceptable to the AEC. A principal difficulty is that steam pressure building up inside an overheating reactor vessel prevents cooling water from rushing in promptly enough to avert melting temperature. Yet, such a system is the only emergency measure that could prevent a meltdown in the event of a catastrophic breakdown in the cooling system. So far, no emergency core-cooling system has even been attempted for the fast breeder, which is cooled by liquid sodium.

In a report on nuclear-reactor safety in July 1971, the Union of Concerned Scientists of Cambridge, Massachusetts, stated that in the event a light-water reactor lost its cooling water, "the reactor core would be expected to melt down and breach all the containment structures, very likely releasing some appreciable fraction of its fission product inventory. The resulting catastrophe and loss of life might well exceed anything this nation has seen in time of peace." This organization and other critics of the AEC's reactor-safety efforts contended that until an emergency core-cooling system can be devised—one that will satisfy all conditions—nuclear-reactor construction should be stopped.

That proposal was made directly to the Select Committee of the Pennsylvania Senate by an AEC scientist, Dr. John W. Gofman, a nuclear chemist and a physician on the staff of the AEC's Lawrence Radiation Laboratory at Livermore, California. A compact man in his fifties, with a predilection for sandals and an Elizabethan beard that lent him the mien of a Renaissance savant, Gofman suggested at the hearing that the Pennsylvania legislature consider a five-year moratorium on the planning, construction, and operation of new atomic-power plants built above ground.

The reasons, he said, were that nuclear electric power has been developed "with the most grave failure of appreciation of the radiation hazard to the population" and that it represents "an antidemocratic disfranchisement" of citizens in several respects. It exposes them to hazards without their consent and also threatens them with property loss without compensation.

During his testimony, Gofman produced a copy of his Homeowners Insurance Policy, issued by the Hartford Insurance Group. The clause entitled "nuclear exclusion" stated: "This policy does not insure against loss by nuclear reactor or nuclear radiation or radioactive contamination, whether controlled or uncontrolled or due to any act or condition incident to any of the foregoing. . . ."

"I urge every Pennsylvanian to examine his home-owner's insurance," Gofman urged, "and ask himself whether he likes the risk to his life, whether he enjoys disfranchisement for an enterprise about which the insurance industry is too skeptical ever to risk dollars."

In its famous report on the consequences of a reactor disaster coded WASH-740, the AEC's Brookhaven National Laboratory estimated that a serious accident in a reactor smaller than the one proposed at Meshoppen could inflict $7 billion in losses in a populous region.

"The individual lucky enough to escape with his life from such an accident," Gofman said, "stands to recover a maximum of seven cents on each dollar lost."

This inadequate amount of nuclear-disaster insurance protection, he explained, is provided under the Price-Anderson Act of 1957 which limits total liability to $582 million for any single nuclear-plant disaster. Through three pools, private insurance companies provide the reactor licensee with $84-million worth of property insurance and $82-million worth of liability insurance for one location. For liability in excess of $82 million, the AEC would indemnify the licensee up to $500 million. In the case of a natural disaster, such as that inflicted on Louisiana and Florida by Hurricane Betsy in 1965, the maximum liability allowable under the Price-Anderson Act would not cover the damage. The Pennsylvania Insurance Commissioner, George F. Reed, testified on this point before the Select Committee. According to one estimate, insurance loss from Hurricane Betsy amounted to $715 million; another estimate cited losses of $1.2 billion in Louisiana alone.

In a disaster to a nuclear-power plant costing $500 million— the estimated cost of the Meshoppen breeder—the allowance of Price-Anderson insurance to the utility and the government's underwriting of the utility's liability appeared grossly inadequate. In citing the insurance situation, Gofman had made a telling point for the citizens' groups. Many of the members had not been aware that private insurance companies would not assume liability for nuclear accidents without government subsidy. It confirmed their worst fears about the nuclear risk.

A Plutonium Economy

Dr. Gofman was emphatic in his testimony about the hazards of the breeder reactor. The individual, he said, is threatened by the plutonium fuel which the fast breeder creates.

> Plutonium in the form of plutonium-oxide particles is one of the most powerful lung-cancer producers known. Release of any plutonium on the surface of the earth irreversibly increases lung-cancer hazards for generations to come—for periods measured in hundreds of thousands of years.

It was unbelievable that serious consideration was being given to an above-ground, fast-breeder reactor in Wyoming County, he continued;

> This is not only a potential disaster for Wyoming County, but for a large part of the eastern seaboard. One millionth of a gram of plutonium is the order of the amount required to produce lung cancer. Any—just any—mishap in handling the ton quantities of plutonium associated with fast-breeder reactors can compromise the future of countless generations of humans.

Wave of the Future

In the fall of 1970, the Meshoppen breeder was being considered by the AEC as one of three breeder-demonstration plants. A total of seventy utilities and utility groups were involved in the over-all program, divided into three groups, each clustered around a manufacturer. Participating in the Meshoppen breeder, with the Pennsylvania Electric Company as proposed operator and Atomics International as constructor, were sixteen other utilities and the

Tennessee Valley Authority. A group consisting of the General Electric Company as constructor and of Consolidated Edison, with six other New York State utilities banded together as the Empire State Atomic Development Associates, as operator, was involved in a second breeder-demonstration project. Another twenty-three power utilities, including Swedish and Swiss interests, were participating in the design study. The third group, clustered around Westinghouse, of twenty-nine electric utilities and the Bonneville Power Administration, was considering a fast-breeder site on the AEC's Hanford reservation near Hanford, Washington. The initial planning contemplated that the Meshoppen breeder would start operating in 1977, the other two in 1979 and 1981, but the order was not determined.

The AEC had determined that the Liquid Metal Fast Breeder Reactor was its highest-priority civilian power program. This made the breeder the highest-priority power device in the United States, since no other agency of government was engaged in any significant development of new sources of power. By reasons of this default, the AEC had become a self-appointed Energy Agency and its energy policy, such as it was, became the national one.

Studies of breeder-reactor construction were going on throughout the power industry as a result of this policy. The electric-utility groups involved in the three breeder-demonstration projects, represented about a quarter of the installed, electrical generating capacity in the United States. These utilities looked ahead to the magic year 2000, when all of their wishful expectations would materialize. According to the Edison Electric Institute, nuclear power would then be producing 57 per cent of all electrical energy in America. By the turn of the twenty-first century, the nation's energy requirements would be six times those of 1970, when nuclear energy was providing barely 1 per cent of the nation's electricity. The breeder was the only economical way the industry knew to capture this fabulous future market.

Radiation Emissions

While the radiation hazards of nuclear processes had been argued for a quarter of a century, it was not until the advent of the Meshoppen breeder proposal that the safety of this type of reactor became a public concern—in spite of the fact that the breeder had been in development for twenty years. On this issue, the Pennsylvania Senate Committee hearings provided a more extensive public forum for debate than had the hearings of the Congressional Joint Committee on Atomic Energy on the environmental effects of producing nuclear electric power the year before. The state hearings called upon scientists who were critics of the AEC's radiation-protection policies for expert testimony; the Congressional hearings ignored them.

In a Joint Committee session, Senator George D. Aiken of Vermont had inquired about "the degree of danger to people living in the vicinity" of a breeder, addressing the query to Milton Shaw, the AEC's forceful, gung-ho director of the Division of Reactor Development and Technology, who had been pushing breeder development hard in the agency's laboratories. Radiating reassurance, Shaw replied: "To me, the degree of danger is such that I will live next door to any one of them with my family." [4] Dr. Gofman, who had another view of the radiation hazard, was not called to these hearings, although he had organized a biomedical program at the Livermore laboratory to evaluate the impact of radioactivity released upon man in the biosphere.

A year later, at the Pennsylvania hearings, Herman Dieckamp, president of Atomics International, echoed Shaw's conviction:

> I can say unequivocally with respect to radiation that our plant is so designed that the radiation exposure to the public from normal operation is negligible. At the site boundary, the exposure is only a fraction of that which comes with frequent air travel, living at high altitudes, having medical or dental X-rays, living in a stone or brick house, or even watching television. . . . Opposition to nuclear

plants on the grounds they present a radiation hazard simply is unfounded.

Of the seventeen commercial power reactors licensed for operation in the United States at that time, he said, none has been responsible for radiation-exposure injuries to the general public.

That contention was challenged by a persistent critic of the AEC, Dr. Ernest J. Sternglass, a radiological physicist at the University of Pittsburgh who also testified at the Pennsylvania hearings—though not before the Joint Committee. He presented data purporting to show a correlation between a rapid rise in radioactive gases released from the Dresden I plant of Commonwealth Edison, southwest of Chicago, and an increase in infant mortality at the same time in the vicinity. Dr. Sternglass cited public-health records in Illinois showing that from 1964 to 1966, the period of the high gas emissions, infant mortality increased 141 per cent in Grundy County, where the reactor is located; 140 per cent in Livingston County, adjacent on the south; and 43 per cent in Kankakee County, adjacent on the southeast. The AEC and the Illinois Department of Public Health branded the Sternglass correlation preposterous. But no other explanation was advanced by either agency to account for sudden infant mortality changes in the three counties near Dresden I.

Another challenge to the AEC-industry position on radiation from nuclear reactors came from Dr. Gofman, who told the Pennsylvania Senate Committee that the AEC's standards of permissible radiation could result in 32,000 cancer deaths a year in the United States. He and Arthur R. Tamplin, an associate at the Livermore laboratory, had called for a tenfold reduction in allowable radiation emissions from nuclear plants.

The AEC radiation standards were defended at the Select Committee hearing by Dr. Lauriston S. Taylor, president of the National Council on Radiation Protection and Measurements, which recommended the standards. Taylor insisted that the standards were perfectly safe. It was unlikely that the limit of permissible radiation would be reached by a reactor, he said, but in the event they were, the most pessimistic estimate of fatalities from the radiation would be only eighty-five persons.[5]

Perceptions might vary about the eighty-five persons. To statisticians, they were an abstraction of little consequence. To critics of radiation policy, they were victims of a propitiatory human sacrifice. The question of radiation safety was to become the most bitterly contested aspect of reactor safety. The Atomic Industrial Establishment took the position that a little radiation was harmless, from a clinical viewpoint. Gofman, Tamplin, and Sternglass maintained that even a little radiation was statistically deadly.

Reducing permissible radiation emissions from nuclear-power plants was not an engineering problem, but an economic one. The emissions could be reduced tenfold if the utilities wanted to spend the money to do so. But according to the AEC's official position, it wasn't necessary. There was, admittedly, some risk in the radiation that escaped up the chimney in the exhaust gases from the plants, or in the water used to cool the condensers, or during periods when the fuel rods were changed. But one must take risks for the benefit of cheap power. From the Establishment's point of view, the benefits far outweighed the risks—except for people living immediately downwind of a big nuclear-power plant.

The Trojan Horse

In the 1970 hearings on atomic-power plants by the Pennsylvania Select Committee—the first public forum on the safety issues in the rising struggle between citizens and the AEC—each side was thoroughly polarized on the questions of plant safety and radiation hazards. Supported by scientist critics of the AEC, notably Gofman, Tamplin, and Sternglass, the concerned citizens regarded the proliferation of nuclear-power plants as hazards of health and property. Supported by the AEC, the Establishment insisted that atomic energy was the key to the future of civilization. The numbers of concerned citizens in the state were small and the organizations tended to act independently of each other. But they were powerful because the questions they raised and the warnings they issued, abetted by the AEC's amazingly clumsy counterpropaganda, worried many thousands of people.

What was happening in Pennsylvania also was occurring spontaneously in a dozen other states in the East, the Midwest, and the West. Citizens in small groups were contesting atomic-power plants all over the country. But the basis of it was opposition, not to technology per se, but to the imposition of a new and hazardous technology without consent and without an acceptable definition of its real dangers. The issue was one of consent.

Beyond the issues of plant safety and radiation hazard, there was a third. Quite apart from the State Senate hearings, it was raised by Representative Dan Flood of Wilkes-Barre, when he attacked the siting of nuclear-power reactors on or near the east coast as a "Trojan Horse" in national defense. They would be easy targets for missiles launched from ships or submarines off the coast. He asserted that a hit would have the effect of detonating a nuclear bomb, emitting huge doses of radiation over a wide area.

The Pennsylvania Congressman cited a section of the AEC's reactor-licensing code stating that an applicant is not required to provide design features or other measures "for the specific purpose of protection against the effects of attacks and destructive acts, including sabotage, by an enemy of the United States." Citing also a letter of September 23, 1970, from the Department of Defense stating that no specific countermeasures are taken to prevent an attack on nuclear plants as such, he said:

> Now, in the face of all this, we learn that Pennsylvania has been selected to have an enormous, first-of-its-kind, experimental, fast-breeder reactor on the Susquehanna River at Meshoppen, thirty-five miles north of my office in Wilkes-Barre. This nuclear experiment is to have as its fissionable core a ton and a half of plutonium. If sabotaged and explosively compressed, which I am assured can be accomplished in a variety of ways, this neighborhood gem will instantly become a huge and incomparably dirty atomic bomb.
>
> The resulting permanent poisoning of the Susquehanna watershed from New York State to the mouth of Chesapeake Bay would probably eliminate the large and heavily populated section of the United States as a habitable region.

The first demonstration Liquid Metal Fast Breeder Reactor did not come to Meshoppen, after all. It was designated for another part of Appalachia, in the forested, mountain wilderness of Tennessee. There, the wave of the future was destined to break late in the 1970s, far from the "madding crowd" of environmentalists.

There was a rebellion against atomic energy in the land. It was a protest unprecedented in the history of technology.

2
A Pattern for Plunder

With the relatively rapid development of the nuclear reactor in only two decades after its invention during World War II, electric utilities began a gradual transition from coal, oil, and gas to atomic fission as a primary source of heat to boil water, make steam, and generate electricity. Out of the new technology of harnessing the heat of atomic fission, a new industry grew. Its development, however, was significantly different from that of the coal industry, which evolved after 1850 when coal replaced wood, and from that of the oil and gas industry, which developed after 1910 to compete with coal. Whereas the fossil-fuel industries were built by private enterprise, the nuclear industry was sired by a government agency, the Atomic Energy Commission, and nurtured by a technology created with public funds. At the outset, this was a logical and acceptable consequence of the government's tight control of all aspects of nuclear technology. Industrial atomic energy could come into being only if the government sponsored it.

Because of this circumstance, civilian atomic power, beginning as a government project, inevitably became a joint enterprise of the agency which controlled access to the technology and the electric-power industry which could apply it to public use. The framers of the Atomic Energy Act of 1954 set up the partnership. "Teamwork between government and industry is the key to optimum progress, efficiency, and economy in this area of atomic endeavor," the Joint Committee on Atomic Energy reported in offering the bill to Congress. Under the Act, the AEC acted as the promoter of the Promethean gift and was also responsible for regulating its use in the public interest. The arrangement implied a strong conflict of interest, in which the public interest was submerged. It is a maxim in Washington that regulatory agencies tend to identify more closely with what they are supposed to regulate than with the public they are supposed to protect. In the regulation of nuclear power, the identification between the AEC and the industry, together with its satellite trade associations, amounted to a commonality of interest and purpose. It was not surprising, therefore, that in the later 1950s there evolved a joint government-industry power structure with common goals and with the access to money and political power to achieve them. I have suggested that this amounts to an Atomic Industrial Establishment devoted to the task of building a fission-reactor power economy, eventually to be based on plutonium fuel, that will last for an indefinite future. In this Establishment the AEC functions as banker and engineer and the Congressional Joint Committee on Atomic Energy as *de facto* board of directors. With the wealth and authority of the government and the investor-owned electric utilities, the Atomic Industrial Establishment is well on the way toward dominating the electric-power industry and shaping the energy future of America.

Monopoly

With the mission of developing atomic energy for industrial use as fast as it could, the AEC found itself obliged at the outset to play nursemaid to an infant nuclear industry. This role was explained in the agency's 1962 Report to President Kennedy. Since the public marketability of nuclear power would depend on economic factors, "the government can best assure widespread use of nuclear energy by fostering developments that make such use economically attractive," the report stated. Hence, the task of the government "includes assuring that the technical developments are carried to the point where, with appropriate encouragement and support, the industry can provide nuclear-power installations of over-all economic attractiveness, sufficient to induce public and investor-owned utilities to install them at their own expense." [1]

This spoon-feeding approach was rationalized in the report with the credo that it was the responsibility of government to ensure the "continuing availability of abundant and economic sources of energy." The AEC's commitment to a philosophy of persuasion in promoting industrial acceptance of atomic energy was defended by the practice of rejecting criticism of the regulations under which the industry was evolving. Disapproval of radiation and safety standards was equated with sabotage of the atomic-energy program; critics were denounced as fools, charlatans, or hysterics by some members or employees of the agency and the Joint Committee on Atomic Energy, as well as by industry spokesmen. The AEC thus acquired a public image of an environmental spoiler, which made it the adversary of a growing public concern about the environment.

A less generally realized effect of this promotional policy was its tendency to allow monopoly to develop in atomic power—a potential foreseen in the debate on the 1954 Act. In a minority report Representatives Melvin Price of Illinois and Chet Holifield of California warned that, "The very magnitude of the economically feasible nuclear-power plants persuades us to believe that the balance will be thrown heavily in favor of private monopoly, unless provision is made for federal development of atomic power, particularly where supply is desired by public or cooperative systems." The bill did not encourage public or cooperative nuclear-power distribution, they said, adding: "Furthermore, it includes no provision assuring that privately owned electric utilities producing nuclear electric energy under license from the Commission shall sell the power at the lowest possible rates, consonant with sound business practice." [2]

In the majority report, the Joint Committee expressed the "firmly held conviction" that increased private participation in atomic-power development "will measurably accelerate our progress toward the day when economic atomic power will be a fact." The majority report said that the bill provided safeguards to prevent special interests from "winning undue advantages at the expense of the national interest."

The antimonopoly restraint in the 1954 Act was a requirement for an antitrust review of the plans of each applicant for a commercial nuclear-reactor license. This restraint, however, was negated by an AEC policy of not issuing any commercial licenses for nuclear-power plants. Of 106 electric-utility nuclear-power plants built, in construction, or planned by the end of 1970, none had even applied for a commercial license. They received or had applied for licenses under the "medical therapy and research development" section of the Act, which did not require a prelicensing, antitrust review. For seventeen years after the Act was passed, the antitrust review was never invoked by the AEC.

The AEC did not issue commercial licenses for reactors which obviously were intended for commercial purposes on the grounds that such a license required a finding that the type of reactor which the applicant wanted to build had "practical value." The determination of "practical value" was a matter of the agency's discretion. The agency steadfastly refused to make such a finding, in spite of the fact that light-water reactors spread rapidly across the country in the 1960s to supply electricity for privately owned utilities from New England to California. In December 1965, the AEC replied to critics of its licensing policy by asserting that: "There has not been sufficient demonstration of the cost of construction and operation of light-water, nuclear electric plants to warrant making a statutory finding that any types of such facilities have been sufficiently developed to be of practical value. . . ."[3]

The Gold Rush

The most persistent critic of the licensing policy on the Joint Committee was Senator Aiken, the Vermont Republican. "The reason for seeking a license for research and development rather than a commercial license is a lesson in transparency," he told the Legislative Conference of the National Rural Electric Cooperative

Association in Washington, D.C., January 15, 1968. "Under a research license, a corporation in effect avoids possible embarrassment under the antitrust laws."

Aiken pointed out that the only way municipal and cooperative power systems could be protected against nuclear monopolistic practices by investory companies under the AEC's antitrust policy was to make complaints after the monopoly had been established. Relief might then be forthcoming, but only after costly, time-consuming litigation.

To most reasonable men, the refusal of the AEC to declare that light-water reactors have practical value (the condition of a commercial license) looked incongruous, he said, "especially when we learn that the cost of producing nuclear power is said to be well under 4 mills per kilowatt hour and the large power companies are falling all over themselves to get licenses." He suggested that while a utility is applying to the AEC for a research license, utility officials are running down the street to the Securities and Exchange Commission to get permission to issue stock in a profit-making enterprise.

It thus appeared that the only "practical value" in reactor-licensing recognized by this policy was the practical value of evading a prelicensing antitrust review. Corporate utility interests were seeking a monopoly over much of the nation to exclude public and cooperative systems before Congress could alter the AEC's policy with corrective legislation, Aiken charged. "The Gold Rush is on," he told the cooperatives.

The Senator from Vermont was not whistling Dixie. Because of this licensing policy, some municipally owned and cooperative power systems found themselves out in the cold when they sought to take part in or buy power from big nuclear-power plants being developed by groups of investor-owned utilities.

The "outsiders" made several notable attempts to protect themselves by forcing the AEC to issue commercial licenses for obviously commercial reactors. They appealed to the courts. In 1969, the AEC's licensing policy was challenged by eleven North Carolina communities and a group of Massachusetts municipal electric sytems. Both groups were shut out of participation in investor utility-financed nuclear-power plants in their respective service areas.

The Yankee Diddle

The North Carolina case arose when the Duke Power Company sought authority to build three large reactors in Oconee County, South Carolina, costing about $314 million. The three reactors would produce 2600 megawatts of electricity—enough to supply one-third of Duke's anticipated customer demand by the time they went into service. The North Carolina cities, which were major customers of Duke, complained of being denied an opportunity to take part in the venture. They protested to the AEC that the denial violated the spirit of the antitrust laws and insisted on getting a piece of the project through their representative, Piedmont Cities Power Supply Company.

A similar complaint was filed with the AEC in the Massachusetts case. Representing forty-two communities, the Power Planning Committee of the Municipal Electric Association of Massachusetts intervened in the licensing hearing on the application of the Vermont Yankee Nuclear Corporation to build a 1665-megawatt nuclear-power plant at Vernon, Vermont, costing $123 million. The municipals asserted that if excluded from buying a share in the proposed plant they would be unable to purchase energy at the same cost as the shareholders of Vermont Yankee, which had been organized for the project by ten New England utilities. Each of them, as stockholders, would receive a contractual supply of power from the reactor.

In both the Duke and Yankee projects, licenses were sought under the medical therapy and research development section of the Act. The developers were thus free of any prelicensing, antitrust compulsion to share the power with the small, city and rural cooperative power systems. The petitioners demanded that the applications made by the utilities should be considered under the commercial-licensing sections of the Act, requiring antitrust review. The AEC denied the petitions. It stuck to the story that it could not issue a commercial license because that required a finding of practical value. No determination that a nuclear reactor had practical value had ever been made. The petitioners then appealed to the United States Court of Appeals in the District of Columbia.

The AEC maintained that a finding of practical value not only presupposed that the nuclear reactor was technically feasible, but also required an economic test. The test was whether the reactor was competitive with fossil-fuel plants. And that still remained to be demonstrated, the AEC argued, for, pending the completion of large atomic-energy plants and the assessment of their operations, there "has not been sufficient demonstration of the cost of construction and operation of light-water nuclear electric plants to warrant making a statutory finding that any types of such facilities have been sufficiently developed to be of practical value."

The agency was upheld. The majority opinion of the three-judge court of appeals noted that on the basis of then-existing data, the AEC was not able to find that the proposed large-scale plants were of practical value. "The Commission apparently hopes that the facilities which are the subject of this action will provide the necessary data for making this determination," the opinion said. "This position not only conforms to the wording of the Act but also adheres to the expressed intent of the drafter of the 1954 law."

A dissent from this opinion was made by Chief Judge David L. Bazelon. On the question of practical value, he commented, the Commission's reasoning "has a nice ring in the abstract but the results are at odds with the facts."

Pointing out that Vermont Yankee had shown the Securities and Exchange Commission that it expected to produce power from the new reactor at a cost lower than that of conventional power plants, Judge Bazelon said:

> These nuclear plants are designed for regular, commercial service. They are financed entirely by private capital without government

subsidy; indeed, private investors have staked nearly half a billion dollars on the commercial success of the plants in these cases alone. The Commission concedes that the basic technology has been proven and that these projects are subject to no more than reasonable business risks. Nonetheless, as a result of the Commission's peculiar standard, these plants are officially still not of practical value. . . .

Thus, in practice, the only substantive consequence of the Commission's reluctance to find practical value has been to excuse the Commission from considering the anti-trust implications of all nuclear projects licensed today. . . . If there are reasons for the Commission's procrastination, the agency has not offered them. . . . Nothing in the Act requires that practical value means absolute certainty of commercial success.

The judge also said that the AEC served as an aggressive and effective salesman for nuclear power, hastening its commercial development. But when nuclear facilities reach the point of practical value, he remarked, the Commission acquires new regulatory duties under the statute.

"The tension between its roles as promoter and potential regulator of nuclear facilities may account for its often irreconcilable statements on their commercial prospects," he added. "We may hope that the agency's duties or its statements will be harmonized before another of these cases comes to court." [4]

Concentration of Control

The Municipal Electric Association of Massachusetts launched another attack on the holding-company squeeze-out of the municipals and cooperatives by petitioning the Securities and Exchange Commission for an evidentiary hearing in order to prove that the exclusion of the municipals from Vermont Yankee was not in the public or consumer interest.

The plight of the municipal electric utilities doing a business of more than $46 million a year in Massachusetts was related by the association's secretary-treasurer, Michael F. Collins, at the hearing of the Joint Committee on June 13, 1968.

The municipals purchase approximately 1.7-billion kilowatts from investor-owned companies at an average of some 12 mills per kilowatt hour, Collins said. To reduce costs, the municipals were seeking opportunities to participate in the ownership of large-scale nuclear generators. It was expected that the Yankee plants planned not only in Vermont but also in Maine could produce power for 4 mills per kilowatt hour, a substantial reduction from what the municipals were paying for power.

"During the last three years," Collins said, "we have been excluded from participating with the New England investor-owned utilities in the currently programed nuclear generator."

The Vermont Yankee consortium pattern was duplicated by the Maine Yankee Atomic Power Company's project, in which eleven private utilities had purchased stock to build a $140-million plant at Wiscasset, Maine. Fearing a lockout in Maine, the Eastern Maine Electric Cooperative, Inc., also petitioned the SEC for an evidentiary hearing.

Both petitions were denied by the SEC and the petitioners appealed. The Court of Appeals, criticizing the SEC for taking too narrow a view of anticompetitive aspects of the cases, remanded them to the securities agency for hearing:

> The assertions of the Municipals as to which they seek an evidentiary hearing point to an increasing concentration in Massachusetts and, indeed, in New England, of control over low-cost electric power through nuclear generating plants. Municipals are excluded from the opportunity to have access on reasonable terms to this power at its source, where it is available solely to the sponsoring companies. Municipals assert that this is due to an intentional course of conduct designed to increase the control of sponsors over the industry in the area and to foreclose opportunities to Municipals.[5]

This pattern of exclusion was not peculiar to New England or the Carolinas. Fred G. Simonton, executive director of the Midwest Electric Consumers Association, Inc., told the Joint Committee at a hearing on June 12, 1968, that: "We can state categorically that our experience has been that with one or two exceptions

... the companies will not voluntarily agree to permit our small systems to participate in joint ownership to share in the savings experienced from the operation of large plants through reduced wholesale rates."

Closing the Loophole

Attempts to close the antitrust loophole in the 1954 Act had gained momentum in the first session of the Ninetieth Congress, when Senators Aiken and Robert F. Kennedy of New York introduced Senate Bill 2565 on October 23, 1967. The bill made a finding of practical value mandatory for the issuance of a reactor-construction license to a utility, thus requiring a prelicensing antitrust investigation. This feature aroused the fierce opposition of some privately owned utilities and of their allies, the equipment manufacturers, but commanded the fervent support of the publicly owned utilities and the cooperatives.

The Aiken-Kennedy bill expressed the proposition that all utilities, regardless of ownership or size, should be able to share in the benefits of nuclear reactors, which were a gift to the utility industry from the taxpayers. Large reactors offered the economy of scale—wherein the larger the plant, the more economically it could produce electric power. But large reactors were expensive, costing hundreds of millions of dollars. Only the largest utilities, or consortiums of utilities, could assemble the capital to build them. The Aiken-Kennedy bill provided a means by which the AEC could be required to examine any antitrust, monopolistic implications of a consortium plan, as a condition of issuing a license.

At the time the bill was introduced, private-investor utilities controlled 84 per cent of the nuclear-plant capacity then operating or on order, 10 per cent was owned by the government, and 6 per cent by municipal and cooperative electric systems. The philosophy of the bill was summarized by Senator Edward M. Kennedy of Massachusetts shortly after the assassination of his brother Robert:

The small, consumer-owned system is the only form of competition faced by the large, private companies on the retail level and is a major factor in keeping down electric utility rates. If the door to nuclear generation is not opened to the smaller utility to secure low-cost bulk power, the result could be complete control of the electric industry by a few, large utilities.[6]

During the debate on commercial licensing, President Nixon's science adviser, Lee A. DuBridge, stated bluntly:

In essence, under existing law, the inherent time lag between economic impact and demonstrated commercial operation shields a significant number of commercial nuclear power plants from the antitrust review prior to construction. The problem is not unique to the current class of light-water reactors, but would be expected to recur as advanced converters and breeders are built in commercial sizes.[7]

The debate about the exclusion of the small utilities from the benefits of nuclear power waxed intermittently before the Joint Committee. During 1968, spokesmen for the have-nots and the haves expounded their views. Some of the direct testimony offers a more coherent picture of the effects of the AEC's licensing policy than voluminous legal documents arising out of complaints against the policy.

Testifying before the Joint Committee on June 13, 1968, Michael F. Collins of the Municipal Electric Association of Massachusetts and its attorney, George Spiegel, related the association's experience in trying to buy power from the new Pilgrim Nuclear Power Plant of the Boston Edison Company.

SENATOR AIKEN: You mentioned . . . that Boston Edison refused to sell power from the new Pilgrim nuclear plant. What reason did they give?
SPIEGEL: They have simply refused and in their last letter they said that the reason they were refusing was, in effect, that they had sold all the nuclear power they had in excess to their needs to three other utilities—Montaup Electric, as I recall; New England Gas and Electric and Northeast Utilities. We have also tried to buy into Maine Yankee and have received the same response—that no, they would not sell to us.

Aiken pointed out to the Joint Committee that New England Gas and Electric and Northeast Utilities are holding companies. Montaup, Spiegel added, is the power-supply company for the Eastern Utilities Associate system, which was one of the three proposing to combine with Boston Edison and the New England Electric System to form the Eastern Electric Energy Association.

> AIKEN: But in selling to the holding companies, it means that you would not know who the ultimate users of the power would be, but you would know that none of the Massachusetts municipals would be included in the list of users. Is that right?
> COLLINS: That is correct, Senator.
> AIKEN: In other words, they would sell to anybody but you.
> SPIEGEL: That's the interpretation we are placing on it.[8]

The advantage of being able to buy power from a large nuclear-power plant was illustrated in reports which attorneys for the Municipal Electric Association of Massachusetts filed with the Joint Committee. They showed that during 1969, the second full year of its operation, the Connecticut Yankee Atomic Power plant was producing electricity at lower cost than competing fossil-fuel plants. Its cost was 5.6 mills per kilowatt hour compared to fossil-fuel energy costs ranging upward from 6.5 mills per kilowatt hour. Energy from Connecticut Yankee was more than 20 per cent below the cost of two oil-fired plants (Canal I and Brayton Point III) placed in service in 1968 and 1969, according to the report.

The response of the investor-owned electrical-utility industry to the reforms offered by the Aiken-Kennedy bill was clear and unequivocal: it was no. Spokesmen for the Edison Electric Institute, a trade association with a membership serving more than three-fourths of the United States, the Consumers Power Company of Michigan, the Southern California Edison Company, and the Wisconsin Power Company complained that the bill would produce unacceptable delays and uncertainties in nuclear-plant licensing. Robert H. Gerdes, president of the Institute, averred that the bill "would open the door to a vast expansion of federal power." The institute "strongly opposes enactment of this legislation," he said.[9]

The AEC firmly supported its investor utility partners. The agency's position was presented to the Joint Committee by Commissioner James T. Ramey, a lawyer who exerted a powerful influence on the agency's policies during this period:

> A basic problem we have with S 2564 [the bill] is that it could have the effect of discrimination between nuclear and conventional power plants . . . the problems of participation by smaller utilities in the benefits of the economies associated with large-scale nuclear facilities exist also in the case of large-scale conventional fuel facilities.

Ramey objected also that the scope of the bill would increase the regulatory responsibilities of the AEC and inject it "into some areas beyond its presently developed competence."

"The Commission's regulatory role has been directed toward matters concerning the protection of the public health and safety from the radiological effects of the operation of nuclear facilities and the common defense and security. . . ." [10]

As I will relate in another chapter, the question of how well the Joint Commission fulfilled its regulatory role to protect public health and safety from radiological effects of nuclear-power plants was being challenged on another front. In this respect, too, it acted as a partner rather than as a regulator of the industry.

Aiken questioned Ramey about a power-plant-siting study which the AEC made in conjunction with the Edison Electric Institute, with no participation by the municipally owned utilities through their trade group, the American Public Power Association, or even by the Federal Power Commission. The existence of the study had been disclosed inadvertently by Milton Shaw, the agency's director of reactor development. Ramey explained that the study with the Institute was merely a supplement to a broader survey that would involve the public-power group. Ramey admitted that the disclosure was unfortunate but said he didn't think "there was any intent that we were just cooperating with the Edison Institute." [11]

Extensive hearings on the Aiken-Kennedy bill were followed by concessions from the AEC and members of the Joint Committee that effectively eliminated a finding of practical value as a condition of commercial licensing. Before the end of the Ninetieth Congress, it had become clear to neutrals as well as to critics in Congress that the AEC's position on "practical value" was simply holding at bay any prelicensing intervention of the antitrust laws in atomic-power-plant development. Had this been the carrot on the stick which persuaded the investor-owned utilities to move toward nuclear power? Whether it had served that purpose, the effects of the policy had been so thoroughly discredited by 1970 that there was no effective opposition to its modification.

The remedy approved by the Joint Committee was the consolidation of several bills which had in common the elimination of the practical-value requirement. It provided that the AEC would make a finding as to whether the plans of the applicant would violate the antitrust laws. The agency could use discretion in inviting an opinion from the Attorney General, unless an intervenor asked that the Justice Department look into a license application. In its report on the bill, the Joint Committee said that it intended any AEC finding of a potential antitrust-law violation be based on "reasonable probability of contravention of the antitrust laws or the policies clearly underlying these laws." The "reform" bill was enacted on December 19, 1970. All new applications would thenceforth be considered under the commercial-licensing section of the Act, subjecting the applicants to antitrust review by the Commission or the Attorney General, except for reactors in the AEC's power-demonstration program. The exception included the forthcoming breeder-reactor demonstration project.

The Joint Committee bill, as a compromise, had been drafted by Senator John O. Pastore of Rhode Island and by Representatives Craig Hosmer of California, Holifield, and Price. It pleased neither side on the monopoly front. Industry spokesmen complained that the bill would add to the delay in getting a reactor-operating license, a process that was becoming protracted by environmental and safety issues as well. The American Public Power Association charged that the compromise actually weakened the antimonopoly provisions of the 1954 Act. It required a finding of "reasonable probability" of antitrust law violation, a more circumspect procedure than the 1954 Act provision which simply called for a showing that a proposed license would tend to create or maintain a situation inconsistent with the antitrust laws. In these complaints, the nature of the compromise was clearly delineated. In exchange for subjecting licensees to the threat of an antitrust review, the compromise bill made such a review less likely than under the old Act—if the practical-value gimmick had not forestalled it.

The Duopoly

In general, the policies of the AEC favored the strong over the weak, for a practical reason: the big boys were the ones who could do the job, even if it had to be done at the expense of the little boys. This philosophy became particularly clear in reactor manufacturing, where the "Big Two" were Westinghouse and General Electric. Both firms had developed the reactor for the submarine *Nautilus,* under the watchful eye of Vice-Admiral Hyman G. Rickover. Westinghouse installed a Pressurized Water Reactor (PWR) of the design developed for the submarine in the first commercial power plant at Shippingport.

From 1946 to 1964, General Electric managed the Hanford laboratories of the AEC at Richland, Washington. The firm gambled that it could sell its Boiling Water Reactor (BWR) to private enterprise without AEC subsidy and won. Between them, Westinghouse and General Electric divided three-fourths of the reactor manufacturing business between them. As of December 1970, a nose count by the Atomic Industrial Forum, the industry association, reported that of 106 reactors operating or planned with capacities of more than 100 megawatts (100,000 kilowatts) net, General Electric had 42; Westinghouse, 34; Babcock and Wilcox, 14; Combustion Engineering, 11, and Gulf General Atomics, 1. Four of the planned projects had not selected a manufacturer. "There can be little doubt that some firms have obtained from their government contracts a commercial advantage in their private nuclear business," observed a study by the Brookings Institution, one of the leading analysts of the nation's industrial economy, of the AEC's contracting practices. "It is no coincidence that the two monarchs of the civilian nuclear power business, Westinghouse and General Electric, have long operated AEC reactor laboratories." [12]

The Brookings study coined the term "Duopoly" in describing the position of General Electric and Westinghouse in the nuclear-reactor field. In its report the study said:

> While the Commission has given lip service to promoting true competition, it has in fact never been willing to give up the strong industrial contractor with a good history of accomplishment. It is obvious, therefore, that the "haves" would become stronger and the "have nots" would become weaker. This situation has given rise to the duopoly which exists today in atomic power.

What Happened to Coal

However questionable the policies of the AEC may have been in favoring the powerful private interests, they worked. Atomic energy arrived in the 1960s as a competitor to fossil fuels and the Atomic Industrial Establishment became a going concern. But even though abetted by these policies, and subsidized to the hilt, the nuclear industry could not have achieved even temporary competitive standing with fossil fuels without changes in its technology. General Electric's boiling-water reactor and Westinghouse's pressurized-water reactor were less efficient than the old-fashioned coal-fired or oil-fired plant, releasing about 10 per cent more waste heat. But the reactors had a notable advantage. The price of fuel for nuclear-power plants was controlled by a friendly AEC, while the price of fossil fuels kept going up. By the end of 1967, nuclear power was in "competitive contention" with fossil fuels, according to the noted power consultant Philip Sporn, whose word is deeply respected by the federal agencies.[13] The exception to this, of course, was the nuclear plant situated relatively close to coal-production centers.

Slowly but surely nuclear energy was acquiring its place in the power market. Between 1965 and 1967, Sporn estimated, 40,000 megawatts of nuclear-power capacity which had been ordered in only twenty months would cause a sales loss to the coal industry of 105 million tons of coal a year. In addition, the coal industry was losing 13 million tons a year to nuclear competition already installed. Yet, in spite of these losses, Sporn noted, coal-industry leaders were considering significant price increases. Some major producers were refusing to sign long-term contracts because they anticipated further price increases. The coal industry failed to realize, Sporn commented, that every time coal prices went up, the competitive position of nuclear power was enhanced. In many parts of the country, he added, the cost difference between nuclear energy and coal has narrowed to where relatively small changes in either direction by one or the other of these competitors can swing the competitive balance.

What did it take to swing the balance? In August 1970, the Council of Economic Advisers reported that bituminous coal prices rose at an annual rate of 56 per cent in the first six months of that year. Residual oil burned by power plants rose at an annual rate of 47 per cent. At that time, the American Public Power Association complained, utilities were receiving a poorer quality of coal at a rising price—when they could get it; some of the smaller utilities feared that they could not obtain adequate fuel supplies for their plants during the spring of 1970.[14]

In the face of a rapidly expanding power demand throughout the country, the emergence of nuclear power did not appear to have any general influence on the coal industry, but the coal industry was having a profound one on nuclear power. Uncertainties of fuel prices and supply were persuading investor-owned electric utilities to go nuclear. And the fossil-fuel industry was lending encouragement to a potential competitor. Producing less than 2 per cent of the nation's power in 1970, however, nuclear power could not be expected to exert any appreciable impact on the fossil-fuel market. A long-term influence was foreseeable, but effective competition was threatened by monopoly trends in fuel extraction and processing. The most conspicuous trend was the rapid growth of the "energy company" with "positions" in coal, oil, and uranium.

This situation was illuminated for the Senate Judiciary Committee's Subcommittee on Antitrust and Monopoly by an analysis of the energy market prepared by the National Economic Research Associates (NERA), an independent consulting firm in Washington. NERA reported that one of the largest owners of coal reserves in the nation was the Humble Oil Company, a domestic affiliate of Standard Oil of New Jersey, which had, for example, bought up a substantial portion of the total coal reserves in Illinois. The Chicago utility market was getting coal from Humble Oil. By 1970, of the 25 largest oil companies in the United States, the NERA study showed, all had positions in natural gas, 18 had positions in oil, 18 had gone into uranium, 11 into coal, and 7 in tar sands. In nuclear energy, 23 per cent of the nation's uranium-milling capacity was owned at the time by an oil company, the Kerr-McGee Corporation, which partly owned another 4 per cent. Eight per cent of the milling capacity was owned by Standard Oil of New Jersey. Moreover, eight of the fifteen uranium discoveries in 1968 were made by oil companies, according to a report by the President's Office of Science and Technology. Oil companies were producing 14 per cent of domestic uranium in 1969 and by January 1, 1970, they controlled 45 per cent of the known uranium reserves.[15] The oil company acquisitions across the energy market spectrum "may be viewed as classic horizontal integration on a scale comparable to the formation of the trusts in the latter decades of the nineteenth century," asserted the NERA study.

The Virtue of Bigness

An economic factor which seemed to favor the transition to nuclear power was economy of scale—the principle that big power plants can sell electricity more cheaply than small ones. Large nuclear-power plants faced less difficulty in getting assured supplies of fuel than big fossil-fuel plants. Bigness had come to the fossil-fuel plants shortly after World War II.[16] In 1948, there were only two power plants in the United States as large as 500 megawatts. By 1960, there were 140 fossil-fuel plants of that size. Plants of 1000 megawatts were being built, plants of 1300 megawatts were on order, and plants of 3000 megawatts were contemplated. But, because of dislocations, monopoly trends, and labor difficulties in the coal industry, the available large reserves of coal that must be blocked out and committed by long-term contract to fuel the giant power plants for a reasonable time were dwindling, especially in the eastern half of the United States where the power demand was growing fast. For example, a 3000-megawatt plant would require a coal reserve of 200 million tons.[17]

A similar apparent shortage was developing in natural gas. NERA characterized it as "artificial." There was plenty of gas in the ground. It simply was not being taken out and piped to customers at an adequate rate, the research organization contended, explaining:

> The producers claim that the price ceilings imposed on them by the Federal Power Commission are too low to provide sufficient incentive to find and develop the gas. . . . The gas producers, which is to say the oil companies, expect that their point of view will prevail with the Federal Power Commission and so have everything to gain by holding off on gas finding and development until they get the price they ask.

Moreover, the fossil-fuel power plants faced a similar situation in oil. NERA showed that the domestic oil output had not been sufficient to meet consumption at prevailing prices. Fuel-oil supplies for utility boilers were inadequate because refineries cut production in favor of more profitable gasoline or distillate oils. Under such circumstances, electric utilities which wanted to realize economy of scale by building giant power plants would find themselves better off with nuclear power. For the time being, a supply of enriched uranium at a controlled price was available. The future? The AEC would take care of it.

The Guaranty

The surety that the fission-power establishment could last for centuries, possibly for millennia, was the breeder reactor. Without it, fission power was doomed—it would fade away with the depletion of uranium-235 in a matter of decades. Fuelwise, the breeder promised a fission-power economy that could meet anticipated expansion into the twenty-first century. The present generation of light-water reactors would be succeeded by a new one of Liquid Metal Fast Breeders burning and creating plutonium from uranium-238. The AEC also had considered a fast reactor that would breed and burn uranium-233 from thorium, but had found the plutonium breeder much more attractive.

As mentioned earlier, the Liquid Metal Fast Breeder had been the AEC's highest reactor-engineering priority since 1962. A cost-benefit analysis made by the agency in 1967–1968 predicted large, direct money benefits from low-cost energy produced by the breeder. Moreover, the device would create new wealth in the form of plutonium fuel. In a speech at Las Vegas on November 19, 1970, Commissioner Clarence E. Larson expressed his optimism in the future of the breeder. He prophesied that a thousand breeders generating a million kilowatts of power each could be operated for 1500 years on uranium available in the United States at $30 a pound. This estimate assumed the existence of a reserve of 1.5 million tons of uranium oxide which could be mined at that price after an estimated supply of 212,000 tons at the 1970 price of $8 a pound was exhausted, circa 1980. Even at $30 a pound, uranium was still a good buy for the highly fuel-efficient breeder reactor. For electric utilities turning away from the problems of getting future fossil-fuels supplies at a predictable price, this prospect seemed encouraging.

The Environmental Bind

Growing concern about air and water pollution, especially in acutely affected metropolitan areas, was expected to accelerate the trend toward nuclear power. Nuclear reactors did not release visible smoke and sulfur oxides. Air-pollution regulations requiring low-sulfur coal and oil, in short supply for utilities, restricted the availability of fuels and hiked their cost. It was only a matter of time until Congress would impose a federal pollution tax on sulfur-oxide emissions. President Richard M. Nixon promised it in his environmental message to Congress on February 8, 1972.

But nuclear reactors were not the answer to the environmentalist's prayer. They released more heat to the environment than did fossil-fuel plants since they were less efficient. Plants using water from adjacent rivers, lakes, or bays to cool their condensers returned it hot to the source. The heated effluent threatened adverse bioenvironmental changes near the plant—an effect called "thermal pollution." The consequences are debatable among limnologists and aquatic biologists. In a 1970 study of thermal discharges from all sources, principally from twenty-seven electric-power plants around Lake Michigan, the AEC's Argonne National Laboratory concluded: ". . . the lake-wide effects of man-made thermal discharges into Lake Michigan are negligible and will continue to be so for the rest of this century." [18] Seven of the power plants considered by the study were nuclear, but only one was in operation at the time. (The heat emissions of those in construction could be calculated.)

The study confirmed the likelihood, however, that localized ecological changes could occur in the lake in the vicinity of hot-water discharges.

Of more concern has been the effect of radioactive gas and particle emission from nuclear-power plants, a problem which has produced controversy among radiation experts themselves about the health effects of low-level radiation. In this debate, environmentalist groups have sided with the critics of the AEC's radiation policies. In the public mind, the invisible threat of radiation emissions and the fear that a power-plant accident might engulf a community with radioactivity were powerful reasons to oppose the proliferation of atomic-power plants.

The concerns of citizens' groups about radiological and thermal pollution tended to neutralize the visible environmental protection advantages of a nuclear-power plant over one using coal or oil. On this score, the Atomic Industrial Establishment was confronted by a determined opposition which began to consider the AEC as a public enemy.

3
The Radiation Heresy

The schizoid role of the Atomic Energy Commission as both patron and regulator of the atomic-power industry was glaringly illustrated in the radiation controversy, which blew up around the nuclear-reactor program in 1969. It was generated principally by three scientists who contended that the radiation standards of the AEC were a menace to public health. Two of them, Ernest J. Sternglass of the University of Pittsburgh and John W. Gofman of the Lawrence Radiation Laboratory, Livermore, had aired their views before the Select Committee of the Pennsylvania Senate in the summer and fall of 1970, when the debate was in full swing, and the third was Gofman's colleague at Livermore, Arthur R. Tamplin.

Although Gofman and Tamplin, who collaborated in their attack on the AEC standards, did not agree with Sternglass's correlation linking emissions from nuclear-power stations and infant mortality in their vicinity, the three critics were united on one point: that the radiation standards of the AEC exposed Americans of all ages to increased risk of cancer, because of the agency's refusal to recognize the carcinogenic effects of low-level radiation.

Sternglass argued that radioactive releases from nuclear-power plants permitted by the AEC standards had increased infant mortality in the United States, as had fallout from nuclear-bomb tests. Gofman and Tamplin made a cold, statistical calculation that the standards considered "safe" by the AEC would result in 32,000 additional deaths from cancer in the United States over and above the natural incidence of the disease. They demanded that the standards of permissible radiation emission be reduced tenfold.

The AEC's response to these attacks on its radiation-protection standards was analogous to the reaction of the Holy Inquisition to the assertions of Galileo Galilei that the earth revolved about the sun. This was heresy. Not only did it put on trial the agency's conflict of interest between its obligation to promote the technology and its duty to protect the public, but it challenged a scientific dogma which accommodated the proliferation of light-water fission reactors. The heresy was intolerable, particularly that of Sternglass whose correlations implied that the radiation policies of the AEC were resulting in widespread infanticide. Hence, it should not be surprising—but only dismaying—that the agency ridiculed its critics, sought to disparage their scientific reputations, and in the cases of Gofman and Tamplin, who were employees of the AEC, attempted to suppress the criticism. I am convinced that the agency would have fired Gofman and Tamplin if it could have done so without arousing the anger of the scientific community. As it was, the action taken against Tamplin by his own laboratory could be interpreted as an effort to persuade him to resign his position.

The agency's near-frantic efforts to show up its critics as fools or charlatans were based not merely on a conviction of scientific righteousness, but on the practical possibility that the claims of the critics would provide ammunition for a growing demand for a halt in nuclear-reactor construction. Demands for a moratorium on new reactors had echoed in the legislative chambers of Pennsylvania and Minnesota. The demand for a tenfold reduction in allowable radiation emissions threatened to require wholesale redesign, higher costs in new plants, and the refitting or "backfitting" of old ones with devices to reduce emissions.

In defending its standards, the AEC also was defending its stewardship of the nuclear-power program. The agency was forced to fall back on a dogma created by national and international committees on radiation protection. It was a body of knowledge supported less by research than by supposition and by the scientific standing of the men who created it. Yet, even they were not sure. Year by year, the standards they recommended were revised, mainly in the direction toward which the Gofman and Tamplin critique pointed.

Reduced to its simplest terms, the AEC's nuclear-reactor licensing policy was based on a standard of permissible radiation emissions which assumes that a little radiation is an acceptable public-health risk in exchange for the benefits of nuclear power. The critics tried to bring to bear evidence showing that the risk was underestimated and was not acceptable when its full dimensions were perceived. Fortunately, the AEC failed to silence them.

Anatomy of a Heresy

Radiation pollution from a reactor is caused by nuclear fission in the core. Each nuclear fission produces two radioactive nuclides plus energy in the form of heat. In a 1000-megawatt reactor, about 10 billion times 10 billion new radioactive nuclides are produced every second. Some of the radionuclides come out of the high smokestacks of nuclear-power plants in gaseous form. Some leak out of fuel elements and migrate into the cooling water, which then carries them into the river, lake, or bay from which the cooling water was drawn. Thus, radioactivity escapes from the plant into the environment—into the air and into the water used to cool the plant condenser.

Boiling-water reactors emit more radiation than pressurized-water reactors. The Liquid Metal Fast Breeder Reactor is being designed to emit less than either of them, the AEC claims. But all reactors emit some radiation. Additional leakage of radionuclides occurs when the fuel elements are changed and taken to a processing plant to be recharged. The processing plants, too, leak radioactivity to their environment.

While it is theoretically possible to design a nuclear reactor with "zero emission" of radioactivity, nobody has done it. The machine would be too costly, and from the AEC's point of view, it would be unnecessary. The AEC takes the position that reactor types now in use are safe enough because the radioactivity they emit is too minute to be considered a health hazard to the population. The critics maintained that the standards under which the reactors could emit some radioactivity are not safe and therefore reactors operating under these standards are likely to be unsafe at times, when they approach the allowable limits.

The amount of radiation the AEC allows nuclear plants to emit is based on recommendations by the now defunct Federal Radiation Council (FRC). This body was established by President Dwight D. Eisenhower in August 1959 to advise him and federal agencies on radiation protection, following the controversy about the effects of bomb-test fallout in the late 1950s and hearings on radiation effects by the Joint Committee. The council was composed of the Secretaries of Health, Education, and Welfare; Defense; Labor; Commerce; and Agriculture; and of the Chairman of the AEC. The small staff of the FRC was headed by Dr. Paul C. Tompkins, who moved to the Environmental Protection Agency when responsibility for environmental radiation standards was transferred to that agency, under President Nixon's Reorganization Plan No. 3 of 1970, and the FRC was disbanded. Tompkins was an old hand in radiation-standard setting. He had served as deputy director of the AEC's Office of Radiation Standards and also as director of research, Bureau of Radiological Health, United States Public Health Service.

During the eleven years of its existence, the FRC relied for guidance on the recommendations of the National Council on Radiation Protection and Measurements. This body, presently consisting of sixty-four physicians and scientists, was formed in 1929 by the National Bureau of Standards and chartered by Congress to work with the International Commission on Radiological Protection. The international body, composed of forty members from fourteen nations, was established in 1928 by the International Congress on Radiology.

Both the National Council on Radiation Protection and Measurements (NCRP) and the International Commission on Radiological Protection (ICRP) agree on standards. Their recommendations have been adopted *in toto* by the Environmental Protection Agency (EPA), which is supposed to set the standards. The AEC implements the standards by setting maximum permissible limits for radiation emissions from nuclear-power plants. And these limits influence the design, and consequently the cost, of nuclear-power plants. The lower the limits, the higher the cost.

Acceptable to Whom?

Throughout the history of both organizations, the approach to standard-setting has become more and more conservative as new knowledge about radiation effects was accumulated. The NCRP has reduced its maximum radiation-exposure recommendations twice since 1947 because of increases in radiation-producing technology and the discovery of hitherto unsuspected radiation effects.

The NCRP-ICRP radiation protection guidelines are based on an assumption that it is an "acceptable risk" for a population to be exposed to 5 rads of radiation over a period of thirty years, or one generation. The term "rad" is an acronym for *r*adiation *a*bsorbed *d*ose—the amount of radiated energy absorbed by animal tissue. The rad equals 100 ergs per gram of tissue.

The maximum permissible dose of radiation per year under this standard is determined by dividing 5 rads by thirty years. This comes out 0.170 rads a year, an amount usually expressed as 170 millirads (or 170 thousandths of a rad) a year.

Consider the fact that a dose of 650 rads is lethal all the time, 400 rads is lethal half the time, and 150 rads only rarely fatal; an exposure of only 0.170 rads over a year's time would appear to be inconsequential. So the AEC believes.

But the assumption of what is "safe" is a relative one. In 1934, the ICRP recommended a maximum permissible dose for radiation workers of 1 rad a week. By 1950, this recommendation was reduced to 0.3 rads, and by 1956, to 0.1 rads per week. In the early years of establishing protection standards, it was assumed there was a threshold below which radiation was harmless. Doses lower than the threshold were called "tolerance" doses because it was assumed that human beings could tolerate them without any ill effects.

By 1954, the ICRP abandoned the concept of a threshold and the term, "tolerance dose," was changed to "maximum permissible dose" or "acceptable dose."

Acceptable to whom? As Dr. Lauriston S. Taylor, the NCRP chairman, had told the Pennsylvania Senate Select Committee, even if the existing radiation-dose limit for population groups from all sources of radiation other than natural background or medical sources were reached, the maximum number of leukemia deaths induced would amount to only eighty-five a year. There was some evidence, he added, that all other malignancies, including cancer of the breast, lung, and thyroid, might be greater than the incidence of leukemia—possibly up to five times as much. But there was some argument about this.

Taylor pointed out that observable radiation effects have been seen only at high doses, 700 times higher than the present maximum permissible dose. And the present maximum of 5 rads per thirty years, or 0.170 rads per year, is 100 times or so higher than the average dose people were getting from man-made radiations in the United States.

Critics of the AEC radiation policy did not agree that radiation effects have been observed only at high-dose rates. They argued that the effects can be revealed by the extrapolation of high-dose effects to low rates and by epidemiological studies linking radiation to changes in mortality rates from cancer and genetic diseases. But statistical evidence itself was so highly controversial that so far as the AEC was concerned it represented no evidence at all.

Only on one point did the AEC and the critics of its radiation standards agree. It was voiced by Taylor in explaining radiation to the Pennsylvania Senate Committee. "The effect of radiation on man could be one of the limiting factors in the development of nuclear power."

That appeared to be the crux of the radiation controversy. Critics of the AEC's radiation standards were not only challenging the standards but threatening to upset the accommodation which allowed the nuclear industry to build up to levels of radiation that could have lethal effects on citizens. In its Publication No. 9, adopted September 15, 1965, the ICRP warned that

> any exposure to radiation may carry some risks for the development of somatic effects, including leukemia and other malignancies, and of hereditary effects. The assumption is made that, down to the lowest levels of dose, the risk of inducing disease or disability increases with the dose accumulated by the individual. This assumption implies that there is no wholly safe dose of radiation . . . any exposure to radiation is assumed to entail a risk of deleterious effects. However, unless man wishes to dispense with activities involving exposures to ionizing radiations, he must recognize that there is a degree of risk and must limit the radiation dose to a level at which the assumed risk is deemed to be acceptable to the individual and to society in view of the benefits of such activities. Such a dose might be called an "acceptable dose" with the same meaning as was implied by "permissible dose."

Again, the question was raised: to whom is the "acceptable dose" acceptable? And who would determine its acceptability? Not the eighty-five persons who would die if the AEC's permissible level of 170 millirads was reached, according to Taylor's estimate. To whom, then?

The question of what risk is acceptable to the American people from nuclear reactors is determined by the AEC. When the radiation controversy broke out in 1969, the risk was pegged to the yearly exposure of 170 millirads, which the AEC considered conservative. This limit was based on the recommendations of the NCRP, the AEC said, following the recommendations of the ICRP. But was it?

In calculating the "acceptable risk" per generation, the ICRP considered sources of radiation that, in addition to the natural background and medical radiation, a person in a developed society might receive in thirty years. These sources included radiation from nuclear reactors, but were not confined to them. Thus, the sources of what might be called technological exposure were: [1]

Occupational	1.0 rad
Exposure of adult workers not engaged directly in radiation work	0.5 rad
Exposure of population at large	2.0 rads
Reserve	1.5 rads
Total	5.0 rads

In this breakdown, the maximum dose of technological radiation to the population from nuclear reactors and other sources affecting large numbers of people, such as fuel-processing plants, would be 2 rads per generation, or 67 millirads per year. That would be two fifths of the AEC's permissible radiation limit.

In this respect, the AEC's claim that its standards were based on the recommendation of highest scientific authority was misleading.

Sole Arbiter

The AEC's powers to regulate radiological discharges to the environment was challenged in 1969 by the State of Minnesota, which attempted to require a nuclear-power plant proposed by the Northern States Power Company to conform to radiation standards of the State Pollution Control Agency. These standards were ten times more rigorous than those of the AEC. As I will relate in more detail in another chapter, the utility contended that the Atomic Energy Act pre-empted to the federal government the authority to regulate radiological hazards and that the state could not act lawfully in this area. Besides, the company complained, the imposition of less permissive standards would greatly increase the cost of the plant, and in some respects such standards could not be met by existing technology. The United States District Court in St. Paul upheld the federal pre-emption of radiation standards. In a paper presented at a nuclear-power symposium at the University of Minnesota, Representative Hosmer of the Joint Committee asserted that it was certainly the legislative intent of

Congress to "occupy the field" in the control of radiation to the exclusion of state regulation.[2]

So it was that the AEC at the outset of the radiation controversy was the sole arbiter of the amount of radiation to which citizens would be subject from nuclear-power plants. It was a policy which implied that the citizen must accept the risk of additional cancer and additional infant deaths as the price of the benefits of nuclear power. The agency's bitter defense of this policy against its critics provided the emotional focus of citizen protest against the proliferation of atomic energy.

A History of Suppression

Since its earliest beginnings, the AEC has been battling critics of its radiation-contamination activities, from nuclear-bomb tests in the atmosphere to the establishment of fuel-processing plants in New York State and in Illinois. The pattern has never changed. When the AEC could, it suppressed its critics, or, at least, attempted to do so. When it could not overtly silence them, the agency assigned people to ridicule and refute their work. To become a critic of this agency was to become its enemy. In its persistent and vehement display of this attitude toward critics, the AEC has been unique among government agencies.

One of the first warnings of danger from low-level radiation came from Hermann J. Muller, who received the Nobel Prize in 1946 for his famous experiments with fruit flies, showing that radiation causes genetic mutations. Muller warned in 1949 that increasing exposure of populations to industrial, medical, and military radioactivity was endangering the heredity of large segments of humanity. When Muller's claims were cited during the atomic-bomb tests by the United States in the 1950s, the AEC insisted that the radiation exposure received by all Americans from nuclear explosions up to 1955 amounted to no more than that from a single chest X-ray. This assertion was misleading in view of the fact that high concentrations of radioisotopes were brought down to the ground by rainstorms in several parts of the country from each of the test series, so that occasionally people in Utah, upper New York State, and elsewhere received higher doses than the average calculated for the total population.

In 1955, the AEC succeeded in suppressing a paper which Muller prepared for an International Conference on the Peaceful Uses of Atomic Energy in Geneva, Switzerland. The agency explained that it deemed Muller's paper unacceptable because it mentioned the bombing of Hiroshima although the conference was devoted to peaceful applications.[3]

Similar warnings of radiation hazards were uttered by another Nobel laureate, Linus Pauling, during the early 1960s. The jeremiads of Pauling and others were directed against nuclear-weapons testing in the atmosphere. There is no doubt that in spite of the AEC's persistence in belittling and refuting the warnings of Muller, Pauling, and others, the respect of the international community for their views greatly facilitated the Partial Nuclear Test Ban Treaty of 1963.

The AEC itself had funded basic research on the biological effects of radiation at its Argonne National Laboratory and the Lawrence Radiation Laboratory, but not all of it was reported promptly, if at all. One instance of this is a study on the effects on laboratory dogs of radioactive strontium which began in 1945 at the Metallurgical Laboratory at the University of Chicago and was continued until the last dog died in 1960, after the laboratory had become the Argonne National Laboratory.

Results of the study have never been fully reported, but a partial report was made in May 1969 to the Hanford Symposium on Radiation Biology of the Fetal and Juvenile Mammal by two Argonne scientists, Miriam P. Finkel and Birute O. Biskis. It dealt with the fate of seven dogs treated with radioactive strontium, five of them in the fetal state through injection of the mother, and two more, twelve to fourteen days after birth.

The five puppies which had received the radiostrontium *in utero,* were born with a concentration of 1.33 millicuries per kilogram of weight. This amount was not acutely lethal, but it caused abnormal skeletal growth. Both dogs treated after birth died of bone cancer, one at the age of two years and the other six years later.

Puppies receiving the radiostrontium *in utero* gained weight very slowly after two weeks and acquired vision later than usual. Compared with untreated puppies starting to walk at two weeks of age, the treated puppies could not stand before their twentieth day.

One of them died at seventy-seven days of bronchopneumonia after exhibiting shortness of breath and loss of appetite on the nineteenth day. The animal, a female, was able to walk a few steps at twenty-four days but shortly after that her legs collapsed. Another puppy, also a female, developed difficulties in the left-front and right-hind legs, which seemed painful. X-rays showed areas of increased density in the metaphyses of the long bones. She suffered a fracture of the right tibia in one such area of higher density at seven weeks and a second fracture two and one-half weeks later. For the next two years, the dog propelled herself on her stomach by pushing with her hind legs. At twenty months, she weighed only 3 kilograms and was almost as broad as she was long. She lost weight suddenly and died at the age of two and one-half years.

Both puppies treated after birth had "controls"—untreated dogs for comparison with the treated animals. One of the treated puppies received 0.52 millicuries of radiostrontium per kilogram of weight at 12 days of age. At 27 days, she began to favor the right hind leg and all joints seemed to be thickened. Two days later, both hind legs were lame. However, the dog was able to stand again in seven weeks and walk two weeks after that. At two months, she weighed 1.97 kilograms while the "control" dog weighed 4.3 kilograms. The "control" was "robust, sleek, and well groomed" compared with the treated dog which was "fragile, ruffled, and unkempt." Bone cancer developed in the treated dog's lower jaw.

The second treated puppy, injected at fourteen days, developed a fracture of the eighth rib at three months and the right radius was broken at six months. These fractures were set and healed. At seven months there were fractures of the left radius and ninth rib, and at seven and one-half months another fracture of the right radius. The dog managed to survive eight and one-half years, however, until it died of bone cancer at the base of the skull.

The study explained that since strontium is chemically similar to calcium, it is deposited in bone areas which are being formed or remolded. The most pronounced effect of radiostrontium exposure of the puppies *in utero* was on their skeletal growth. Those bones which were ossifying (forming) at the time of treatment received large deposits of radiostrontium and most of that tissue was killed. Of the five animals treated *in utero,* the longest lived lasted two and one-half years and the shortest, only eighteen days.

The most controversial paper presented at the symposium claimed a connection between the fallout of radioactive strontium (Sr-89 and Sr-90) from bomb testing and the trends of the infant-mortality rate in the United States. It was the work of Sternglass, who was mentioned in earlier chapters for his efforts to focus national attention on the hazards of low-level radiation.[4]

Sternglass was the first to attempt to use an epidemiological method to show that low-level radiation, on the order of the AEC maximum permissible emission levels, could effect the sensitive human fetus. In his correlation, he assumed a cause-and-effect relationship between two phenomena—a leveling off after 1950 of the hitherto steady decline in infant-mortality rates in many parts of the United States and the advent in that period of radioactive fallout, particularly radiostrontium, from nuclear-weapons testing. Strontium-90, the long-lived radioisotope carried down from the high atmosphere by rain or snow, lodges on grass or grain eaten by livestock and thus passes into the human food chain.

The idea that such a correlation might exist occurred to Sternglass as a result of two pieces of data. One was a report on a heavy rainout of radioactive debris in the Albany-Troy-Schenectady area following a 50-kiloton bomb test in 1953 in Nevada, and the other was a report from American and British studies that the children of mothers receiving pelvic X-rays during pregnancy had a significantly higher incidence of childhood leukemia than children of mothers who had not been X-rayed while pregnant. Putting these data together, Sternglass perceived an analogous link between the low-level radiation of fallout and infant-mortality rates.

A medium-sized, dark-haired man in his early forties, Sternglass has pursued this thesis at a high pitch of excitement for nearly ten years. With each new significant finding, he sees with elation the picture growing clearer. "Look at this," he tells his colleagues. "Look what we've found! See how it adds up?" He is like a prosecuting attorney who keeps on digging up new evidence to prove his case, only to be compelled to present it before an indifferent or hostile jury which rejects his conclusions, ridicules his methods, and exaggerates occasional errors in his work. "They won't believe this," he will tell an associate about some supporting data he has found. And they—the AEC and many of the more conservative members of the scientific community—don't. Sternglass is one of the most vigorously refuted and generally maligned critics that the AEC has ever had—and the most influential. The agency has spent many times more man hours trying to disprove his correlation than Sternglass has spent assembling his materials. But in spite of criticism, he goes on to build a case showing that the effect of low-level radiation upon man can only be perceived by epidemiological studies. And it is a significant effect.

Sternglass's academic and career credentials are as good as those of his opponents. He received a doctoral degree in engineering physics at Cornell University, where he is remembered for a controversial theory he proposed on nuclear structure. He worked as a research physicist with the United States Naval Ordnance Laboratory, specializing in electronic instruments, and then spent fifteen years with the Westinghouse Research Laboratory in Pittsburgh, where he helped develop the television camera that astronauts have used on the moon. Since 1967, he has been professor of radiation physics and director of the Radiological Physics and Engineering Division of the University of Pittsburgh.

Sternglass had been impressed by the warnings of Muller and Pauling that low levels of radiation endangered the genetic health of the human species. But the lack of data made these speculations unconvincing and served the ends of enthusiastic nuclear technologists who demanded proof. Industrial atomic energy was developed on the principle that low levels of radiation are harmless. It is a principle that resulted from the same kind of thinking before the Partial Nuclear Test Ban Treaty of 1963 that radiation from tests was harmless as long as it produced no clinical effects.

The Invisible Bullet

Low-level radiation effects could not be perceived immediately. The effects were invisible. They would remain invisible for years until they appeared as a form of cancer. By then the cause was too far away in time to be associated with the effect. But there was a mechanism linking them. It could be visualized if one thought of a radiation particle as a bullet which scored a single hit on a molecule of deoxyribonucleic acid (DNA) governing the reproductive machinery of a cell. If the genetic code contained in the DNA was altered or damaged by a particle of radiation, the vital mechanism controlling the growth, function, and reproduction of the cell might break down. The cell might then begin uninhibited growth, proliferating into cancer. If just one particle of radiation could start this process, which might develop over a number of years before symptoms became noticeable, how could anyone say that even a little radiation was harmless?

In 1962, the work of Dr. Brian MacMahon at the Harvard School of Public Health provided some evidence that low-level radiation could be a significant health hazard. MacMahon showed an increased incidence up to 40 per cent in leukemia among children whose mothers had received diagnostic X-rays during pregnancies, compared with children not exposed. The Harvard study covered 400,000 infants. The X-ray dose amounted to 1 or 2 rems.[5] The term "rem" is another acronym, like rad, meaning "*r*oentgen *e*quivalent *m*an." It is the quantity of radiation which, when absorbed by an organism, produces an effect equal to the absorption of 1 roentgen of gamma radiation. For X-rays, 1 rem (or 1 roentgen) equals 1 rad.

The Harvard study confirmed earlier work published in England in 1958 by Dr. Alice Stewart, head of the Department of Preventive Medicine at the University of Oxford. She had found an increase of nearly 90 per cent in the incidence of leukemia and other malignancies in children exposed *in utero* to X-rays, compared with the incidence of cancer among unexposed children. The lower risk found by MacMahon may reflect the reduction in exposure with improvements in X-ray technology during the interval between the two studies. A later survey by Stewart and G. W. Kneale published in 1968 reported generally a 50-per-cent increase in leukemia and other cancers in children exposed to X-rays *in utero,* compared with children not exposed during that critical, developmental period.

Another study, reported in 1970 by three Johns Hopkins University researchers, investigated the effects of X-ray exposure on 1500 human female fetuses who survived to become mothers, as measured by their reproductive performance and the fate of their offspring. This group was compared with 1500 mothers who had not been exposed to X-rays during their fetal life. The study found that the mothers who had been exposed before reaching thirty weeks of fetal life gave birth to nearly twice as many male as female babies, compared to "control" mothers. Mothers who had been exposed at thirty to thirty-four weeks or at forty or more weeks produced offspring with the opposite sex ratio—more female than male babies—and those exposed at thirty-five to thirty-nine weeks produced an equal number of males and females. The researchers speculated that the sex ratios were affected by chromosome alterations during certain critical periods of fetal development. In the group of mothers exposed before thirty weeks of life, more had precipitate labor and their babies had a longer period of gestation. In general, the mothers who had been exposed as fetuses had more pregnancies, more births, and larger families than the controls. Infant death rates and the rates of twinning, however, were similar among the children of both exposed and control groups.[6]

The Radiation Heresy / 63

Taken together, these studies suggested that the consequences of low-level radiation—less than 5 rads—to the developing fetus were more serious and diverse than the federal radiation standards implied. On the basis of the MacMahon and the earlier Stewart studies, Sternglass published a paper in the June 7, 1963, issue of *Science,* the journal of the American Association for the Advancement of Science, stating that prenatal X-ray effects implied similar ones in unborn children as a result of radioactive fallout ingested by the mother.

Sternglass assumed that natural background radiation triggered 5 to 10 per cent of all childhood cancer and leukemia. The injection of radioactive debris from bomb testing into the environment increased the total radiation exposure of every individual in the northern hemisphere by the equivalent of one to two weeks of natural background radiation. Consequently, Sternglass was certain that the fallout from bomb testing had increased the incidence of cancer, especially among children exposed to the radiation during prenatal life. He estimated that a series of nuclear-bomb tests in the atmosphere, comparable to the combined United States-Soviet Union tests of 1961–1962, would produce additional radiation exposure for every pregnant woman in the northern hemisphere equaling the dose to the embryo of a typical pelvic X-ray taken with the best of modern X-ray machines.

Sternglass went on to suppose that larger doses could be received from bomb tests during a rainout from a radioactive cloud passing over an area. Such a rainout of a radioactive cloud had occurred in the Troy-Albany-Schenectady area of New York State in April 1953. If it had any effects on infants before or after birth, these might show up, he reasoned, in the vital statistics of the period. Reviewing the vital statistics of the State of New York, he suddenly came to the belief that he had struck gold. They did seem to show a correlation between the fallout of radioactive debris—principally strontium-90—and infant-mortality rates in the Troy-Albany-Schenectady area. What he found was that the decline of the infant-mortality rate in the area during the

1935–1950 period had leveled off after the rainout of April 26, 1953. The measured external-radiation dose was 0.1 rads. Sternglass correlated this dose with a slowing of the decline in infant mortality that persisted until 1966.

Five years after the rainout, there was a doubling in childhood leukemia cases for a period of eight years, he reported. He attributed the delayed response in children who had not yet been conceived when the rainout came to genetic damage in the parents.

Enlarging the arena of his investigation to cover all of New York State, Sternglass then said that the fetal death rate began to deviate from the 1935–1950 rate of decline as early as 1951, when the AEC began atomic-weapons testing at the Nevada Test Site. The rate of decline leveled off at about 23 deaths per 1000 live births between 1957–1963 and then the death rate rose to 27.3 in 1964, compared with the national average of 24.8. Sternglass attributed the jump to bomb testing, which was not finally halted in the atmosphere until the Partial Nuclear Test Ban Treaty went into effect at the end of 1963. The 1964 rate was interpreted as reflecting the accumulated effects of tests up to the cessation of atmospheric testing by the United States and the USSR under the treaty. Then, in 1965–1966, infant mortality resumed its old rate of decline, which Sternglass believed was a reflection of the ban on atmosphere testing. Subsequently, he has cited a 1967 report by Dr. Helen C. Chase of the United States Office of Health Statistics Analysis showing that by 1964, when the United States infant-death rate was 24.8, sixteen other nations had achieved a lower rate. They were: the Netherlands, 14.8; Norway, 16.4; Finland, 17; Iceland, 17.7; Denmark, 18.7; Switzerland, 19; New Zealand, 19.1; Australia, 19.1; England and Wales, 19.9; Japan, 20.4; Czechoslovakia, 21.2; the Ukraine, 22; France, 23.3; Taiwan, 23.9; Scotland, 24; and Canada, 24.7. "Among all these nations, it was the United States which was exposed to the most intense fallout from the tests in Nevada and the South Pacific. As recently as 1947, the United States had a lower infant mortality than the Netherlands." [7]

A Wind of Death

In contrast to New York State, the fetal death rate in California maintained a steady decline in the early 1950s because, Sternglass said, California is upwind from the Nevada Test Site. Radioactive debris from the tests were swept eastward by the prevailing winds. However, two to three years after hydrogen-bomb testing began in the Pacific Ocean in 1954, the rate of decline began to slow down in California, too, with the eastward drift of fallout from the high atmosphere.

Sternglass asserted that the changes in fetal death rate in New York State could be blamed on the deposition of radiostrontium on grass and crops, where it would be taken up into the food chain. The statistical correspondence he found in that area between testing and the fluctuation of fetal death rates held true also for the United States as a whole, for all periods of gestation up to nine months, and the infant-mortality rates followed a similar pattern. Increases appeared in four northern, metropolitan states where there had been a heavy rainout after the 1951 Nevada tests. Rural "wet" states in the South, where radioactive dust also had been brought down to the ground by rainfall, showed a similar effect in infant- and fetal-mortality rates.

However, the "dry" rural states in the West, especially in New Mexico, do not show effects of the Nevada tests, Sternglass said, because they did not lie in the fallout path. Only when debris from the large hydrogen-weapons tests was injected into the stratosphere did a leveling off in the decline in the infant-mortality rate show up in those states. It appeared first in the mountain states of Idaho and Colorado in 1954 and later in Wyoming and New Mexico in 1958.

As I mentioned earlier in this chapter, Linus Pauling was among the first to consider the effects of bomb testing in the atmosphere on infants. He had estimated that fission products from nuclear-bomb testing in the atmosphere would produce about 10,000 children with gross physical or mental defects and 100,000 embryonic, neonatal, and childhood deaths. These estimates were uncertain, Pauling said, because of deficiencies in the knowledge about the effects of low-level radiation. "The uncertainty is usually expressed by saying the actual numbers may be only one-fifth as great or maybe five times as great as the estimates, but the errors may be even larger than this," said the Nobel laureate, honored with the prize twice for his contributions to peace and in chemistry.[8]

There is no question that the warnings by Muller, Pauling, and others about the effects of injecting radioactive particles into the biosphere had influenced the decision of both the United States and the Soviet Union to end testing in the atmosphere. After the treaty, the danger of radiation, at least from fallout, had seemed to recede from the forefront of public consciousness until it was revived by the controversy over the deployment of antiballistic missiles in 1968–1969. Both the United States long-range Spartan and short-range Sprint antimissiles were designed to intercept and destroy an incoming warhead just above or in the atmosphere of the territory being defended. The resulting radioactivity drifting downward after interception could have serious, if not disastrous, public-health consequences.

Attempting to add statistical evidence to the earlier warnings of Muller and Pauling, Sternglass made a report of his findings of a correlation between fallout and infant-fetal mortality to the Health Physics Society meeting in Denver in June 1968. The press seized on the story and circulated it widely. Some of the members of the society were alarmed by the notoriety. Sternglass relates that one of them complained in a letter to the society's board that the publicity about his report was damaging to the nuclear industry.

Sternglass sent a paper detailing his findings to *Science,* but it was rejected. He then sent a copy of an article entitled, "Infant Mortality and Nuclear Testing," to the *Bulletin of the Atomic Scientists* in the late fall of 1968, when as managing editor of the *Bulletin* I felt that the paper made an important contribution to the revived discussion on the general effects of fallout and of low-level radiation. Despite strong differences of opinion among members of the *Bulletin*'s editorial board and other experts about the scientific adequacy of Sternglass's argument, I ran the article in the April 1969 issue of the *Bulletin,* launching a year of controversy between Sternglass and his critics. A number of friends of the *Bulletin* regretted the publication of the article because they did not believe it was scientifically sound, but I have always been reluctant to accept such a judgment from one scientist about the work of another, especially when the work is new and controversial. Who shall say what is scientifically sound? Too many advances in science and medicine have had to run the gantlet of this kind of judgment before they were accepted.

In May, following the debut of the Sternglass correlation in the *Bulletin of the Atomic Scientists,* the indefatigable investigator presented a more detailed paper on the correlations between strontium-90 fallout and infant mortality, at the symposium at Hanford, Washington. The paper was roundly buffeted by rebuttal during and after the meeting, but it is one of the few presentations that have made that meeting memorable. The correlation was vulnerable because it amounted to a hypothesis linking two events, without clinical evidence. Other explanations could be surmised for the fetal- and infant-mortality changes which Sternglass attributed to fallout. Critics questioned his statistical conclusions. But no one was able to present a different explanation that would account for the changes in mortality rates and relegate their match with fallout to coincidence.

The controversy over the correlation quickly spilled over into the press. The popular British weekly science magazine *New Scientist* featured an article by Sternglass, entitled, "Has Nuclear Testing Caused Infant Deaths?" Along with it, the magazine ran a sharp reply by Dr. Stewart. She disagreed with Sternglass's assumption that cancers would arise in children as a result of the irradiation of their parents before conception through genetic damage. She generally dismissed the correlation as not proved.

Undaunted, Sternglass continued to press for the dissemination of his theories. He sent a letter to *The New York Times* which attracted the attention of Harold Hayes, editor of *Esquire* magazine. Hayes invited Sternglass to write an article for *Esquire,* and the physicist obliged. The result was a sensationally titled exposé, "The Death of All Children." Its appearance in the magazine was heralded by full-page advertisements in *The New York Times.* In the article Sternglass predicted that the fallout from antiballistic missile detonations in the atmosphere could extinguish humanity.

Suddenly, Sternglass was projected onto the public scene as a new prophet of doom, forecasting the coming of a nuclear Armageddon at a time when public concern about the environment and its contamination by man was blooming. His impact was dulled only by the fact that he had plenty of competition in the doomsday field.

Peak exposure of the Sternglass correlation came on radio and television and in the newspapers as the Senate debated the deployment of President Nixon's Safeguard Antiballistic Missile System. Whether it had any influence on that debate is conjectural, but some members of Congress were impressed. Representatives Jonathan B. Bingham of New York and Lucien N. Nedzi of Michigan noted in the Congressional Record that, in their view:

> the most significant inference that must be drawn from the Sternglass work is its implication for the "first strike" argument that is so widely used to justify proposed new nuclear-weapons systems like the ABM and MIRV [multiple independently targetable reentry vehicle].... If the Sternglass findings are correct, that even low levels of strontium-90 may have devastating effects on subsequent generations of all nations in the same hemisphere where a nuclear first strike occurs, such a first strike becomes unthinkable.[9]

Bingham and Nedzi invited Sternglass to take part in a seminar during the summer of 1969 with two other scientists, Dr. Shields Warren, professor emeritus of pathology of the Harvard University Medical School, and Dr. George Hutchison, an epidemiologist of the Michael Reese Hospital in Chicago. Members of the Joint Committee on Atomic Energy including Holifield, then chairman, attended.

Bingham reported:

Several things became evident in the course of our discussion. First, the correlation Dr. Sternglass has discovered between the presence of radioactive strontium-90 from nuclear tests and infant mortality in the United States appears to be the only explanation currently available to explain the excess infant mortality in this country noted in recent years by the Public Health Service. As a result of this excess infant mortality, the United States has dropped from second or third from the top of the list of the developed nations with regard to low rates of infant mortality to eighteenth on that list. Despite many efforts to find an explanation for this phenomenon, various possible causes have been eliminated . . . and no theory currently has much evidence to support it other than that now offered by Dr. Sternglass.

The Indictment

The reaction from Establishment sources gathered steam after the Sternglass correlation was sensationalized in the mass media. Much of this was directed at what he had indicated indirectly, rather than at what he intended to show. He intended to add to the evidence that there existed no threshold below which radiation does not produce both somatic and genetic effects in man. He intended to show that for the sensitive embryo, fetus, and infant, even relatively low-level doses from peacetime fallout could lead to detectable increases in death rates when data on very large population groups are examined.

What he succeeded in doing, however, was to present to the public an indictment charging that the military and industrial atomic policies of the United States and of other nations were killing babies. Such a charge was intolerable, not only to the AEC, the ABM lobby, and their industrial partners, but also to the entire public-health apparatus which stood accused, by implication, of negligence or misfeasance. For if Sternglass was right, the government of the United States and the public-health apparatus of every state had been derelict to the point of permitting infanticide.

The response to Sternglass was a prompt denunciation of his theories by the agencies which felt themselves accused, namely, the AEC and the Department of Health, Education, and Welfare.

Sternglass's correlation is probably the most widely publicized and passionately refuted theory since the radiation controversy began. Yet, a curious ambivalence appears in much of the criticism of it. While many critics insist Sternglass is utterly wrong on every count, some seem to be fascinated, in spite of themselves, by the possibilities he suggests. The United States Public Health Service rebuttal, a pompous and self-conscious document entitled, "Evaluation of a Possible Causal Relationship between Fallout Deposition of Strontium-90 and Infant and Fetal Mortality Trends," remarks:

> Although all the evidence which Dr. Sternglass has presented to support an association between strontium-90 deposition and a decrease in the rate of decline of infant and fetal mortality in the United States has failed to stand up under careful scrutiny, the important implications of such an association, if true, warrant some further investigation.

After some discussion of the difficulties to be encountered in attempting to establish the association, however, the Public Health Service critique went on to say: "The rate of decline of infant mortality in the United States did change around 1950. This lowering of the rate of decline has been a concern of many people working in the public-health field."

"Stripped of the emotional and political arguments utilized by Professor Sternglass to focus attention on his hypothesis," editorialized the *American Journal of Public Health,* "the facts appear to present a case for further investigation." [10]

The AEC, however, offered no quarter. "None of Sternglass's arguments holds up under examination," said the AEC staff evaluation. "In each case, assumptions are contrary to fact and are based on misinterpretation, distortion, and biased selection of available data."

But, as Representative Bingham observed during his seminar, the Sternglass correlation "appears to be the only explanation currently available."

The Sternglass correlation disturbed a number of scientists. Freeman J. Dyson of the Institute for Advanced Study at Princeton, remarked: Sternglass displays evidence that the effect of fallout in killing babies is about a hundred times greater than has been generally supposed. The evidence is not sufficient to prove that Sternglass is right. The essential point is that Sternglass may be right.[11]

Several others appeared to be impressed more by Sternglass's intuition than by his data. Dr. Karl Z. Morgan, director of the Health-Physics Division of the Oak Ridge National Laboratory, termed the Sternglass correlation highly suggestive, but said that he doubted seriously that Sternglass was right. Nevertheless, he said, the correlation demanded an answer. René Dubos of Rockefeller University was dubious about the correlation, too, but told *Science* it would be "a social crime" not to investigate fully whether fallout has helped cause infant-mortality rates to increase.[12]

Sternglass received support from Dr. Robert C. Pendleton, director of the Radiological Health Department at the University of Utah, who advised him:

> I have followed your problems with great interest and sympathy. I have been through the same kind of vituperative attack that you have been experiencing over the last three years, so I can sympathize from a position of knowledge. Do not let them force you to deviate and simply recognize that opponents of this point of view are worried about the future of their empires rather than about the welfare of mankind.[13]

Pendleton had been criticized by the AEC for calling attention to high levels of radioiodine in milk from the fallout of bomb testing in Nevada.

The AEC marshaled a "Truth Squad" after Sternglass discussed his correlation on the NBC "Today" show. Two proponents of the AEC view of the matter, Dr. John Storer, scientific director for Pathology and Immunology, Biological Division, Oak Ridge National Laboratory, and Dr. Leonard Sagan, a former employee of the division, received equal time on July 28, 1969, to refute Sternglass.

Sagan, an associate director of the Department of Environmental Health, Palo Alto, California Medical Clinic, had served with the Atomic Bomb Casualty Commission of the National Academy of Sciences. In response to a question from the interviewer, Hugh Downs, Sagan asserted: "I think scientists now almost universally agree that the levels of radiation to which the American public was exposed from the fallout source have been harmless."

And in answering a subsequent question from Barbara Walters, Storer confirmed this view. "At high doses, there is no question but that radiation can be damaging," he said. "It can produce leukemia and cancer, but these levels that we are talking about that occurred with fallout are really quite trivial. There is no evidence that they produced any of those effects."

Sagan criticized Sternglass for failing to describe the size and age distribution of the population at risk when he cited increases in leukemia in the Troy-Albany-Schenectady area.[14] He asserted further that the radiation resulting from the 1953 rainout in the area was estimated at 55 millirads. "In a population the size of that of Troy-Albany, New York, based on conservative estimates of the risk of radiation leukemogenesis, at least a thousand times more radiation [or about 55 rads] would be required to produce one additional case per year."

Sagan also attacked Sternglass's assumption that infant-mortality rates would continue to fall on a semilogarithmic decline toward zero. "This assumption is not based on any biological evidence, human experience, or even reasonable judgment," he said. He added that studies of atomic-bomb survivors at Hiroshima failed to show increased mortality among their children. (Sternglass later responded to this by saying there was very little fallout of strontium-90 on the target cities of Hiroshima and Nagasaki.)

Sternglass noted that the AEC had never carried out population studies which might reveal the effects of low-level radiation on a population. Why not? He suggested one reason why this should be done. Within the next thirty years, more than one-half the electrical energy of the United States might be generated by nuclear reactors, "with a truly staggering increase in the amounts of fission products that must be released into the atmosphere."[15]

Reactors and Infant Death

From the relation of fallout to infant mortality, Sternglass moved on to a new correlation between reactors and infant mortality. At this stage in the development of his thesis that man-made radiation exacts an infant sacrifice from society, Sternglass was having difficulty finding journals to publish his papers. In various styles, ranging from the didactic to the emotional, most of the members of the scientific community who commented on his work promptly disparaged it by disputing his data and condemning his conclusions. There were exceptions, as we shall see presently. However, for three years the Sternglass correlation between low-level radiation and infant mortality was subjected to much the same quality of skepticism, criticism, and rejection by the leaders of science as the correlation between smoking and lung cancer had been from organized medicine in the early 1950s. Nevertheless, Sternglass's disclosures continued to intrigue some of the mass media, and the Pittsburgh physicist was in demand as an expert witness for environmentalist intervenors at AEC reactor-licensing hearings. In this manner, at least, he received a public hearing.

In his testimony before the Pennsylvania Senate Select Committee, he had tried to show that increases in the radioactive discharges from the Dresden I nuclear plant, southwest of Chicago, had resulted in an average increase of 48 per cent in infant mortality in six nearby counties during the period 1964 to 1966. As mentioned earlier, Grundy County, where the reactor is located, and Livingston County to the south of it, showed increases of 141 and 140 per cent respectively in infant deaths during that period. Additionally, increases of 43 per cent appeared in Kankakee County and of 5 per cent in Will County on the east. However, the infant-death rates dropped 7 per cent in LaSalle County on the west and 31 per cent in Kendall County to the north.

At first, Sternglass tried to explain the rise in infant mortality in some counties and not in others with the observation that the prevailing wind blows from the northwest. This notion is supported by the existence of the great sand dunes on the eastern and southeastern shores of Lake Michigan, indicating that prevailing westerly and northwesterly winds have been blowing across northern Illinois for centuries. But detailed meteorological records of wind direction to the southwest of the great lake did not support that conclusion in Grundy County in 1964–1966. The data showed that at some times of the year, the wind blows mostly from the southwest, moving radioactive gases toward Chicago.

The AEC was thus able to refute Sternglass's downwind theory, but offered no alternate explanation for the relationship between the rise of radioactive waste emissions from the Dresden I plant and of infant mortality in four counties. These effects were unrelated and coincidental, the agency insisted. The director of the Illinois Department of Health, Dr. Franklin Yoder, issued a statement asserting, "It must be recognized that the number of infant deaths in question from all causes are insufficiently large to document a statistically significant variation." He maintained that there had been "no significant radioactive pollution of the ground, air, or surface waters in Grundy or Kankakee Counties."

When the actual numbers are considered, the increases in Grundy and Livingston counties suggest, circumstantially, that something in the environment was killing babies. Sternglass believed he knew what it was; but could he prove it? All he could show were two discrete events which he linked as cause and effect.

First, between 1964 and 1966, radioactive gaseous emissions from Dresden I rose from 521,000 to 736,000 curies a year and radioactive liquid releases from 3.82 to 11.5 curies a year.[16] The liquid releases are considered more significant threats to public health in spite of the fact they are smaller in terms of curies, because the strontium-90, cesium-137 and iodine-131 that they contain are more likely to be concentrated in the food chain in the vicinity and are much more toxic than much of the gaseous effluent consisting of noble gases. The increase in liquid releases amounted to a rise from 11.6 to 34.8 per cent of the AEC safety limit. Gaseous releases rose from 2.2 to 3.5 per cent of the limit of 22 million curies for noble and activation gases.

The second event was the increase in Grundy and Livingston counties from 1964 to 1966, as shown by Illinois public-health records:

	1964		rate per 1000 births	1966		rate per 1000 births
	deaths	births		deaths	births	
Grundy	7	442	15.8	18	474	38.0
Livingston	6	728	8.2	12	608	19.7

In more populous Kankakee County, the infant-death rate rose from 20.7 to 29.5 per 1000 live births and in even more populous Will County, from 22.2 to 23.3 per 1000. Yoder offered no explanation of his own for these increases, but assured the citizens they were not significant.

Sternglass's Dresden I correlation was vulnerable because there was no clinical evidence to support it. The relationship of the two events could only be assumed. Other undetermined factors could have been responsible for the rise in infant deaths. That was the official position of the AEC and the State of Illinois. It was curious, indeed, that such an abrupt change in the infant mortality rates of four counties near an atomic power plant did not excite public concern. But it did not.

Perhaps, this indifference is a result of the difficulty the layman has in perceiving how an increase in radioactivity might affect infant mortality.

The term "curie" refers to the quantity of radioactive material emitting 3.7×10^{10} (37 billion) disintegrations a second. The effect of these disintegrations on living tissue is measured in terms of rads—the dose that the tissue absorbs. The toxicity of a curie of a radioisotope depends on how readily it is taken up by the vital organs. Two investigators of the Vanderbilt University School of Medicine in Nashville, Tennessee, reported in 1969 that one-thousandth of a curie (1 millicurie) of iodine-131 absorbed by a fetus from the mother in the ninth to twenty-second week of gestation would give a *total* dose of 0.796 to 3.0 rads. If concentrated entirely in the fetus's thyroid gland, however, the millicurie of radioiodine would give a dose of from 715 to 5900 rads—an enormous one.[17]

Stewart and Kneale showed in their 1970 report, mentioned earlier, that exposure of the fetus in the first three months to 80 to 100 millirads (0.08 to 0.10 rads) from an X-ray of the mother doubles the incidence of cancer in childhood.[18] Yet, from just 1 millicurie of radioiodine, the fetus would receive from eight to thirty times the X-ray doubling dose, based on the results of the Vanderbilt study. The pathway of the millicurie of radioiodine is easy to visualize: from the smokestack or liquid discharge of the nuclear-power plant, to the grass or water intake of the cow, and on into the mother from the cow's milk, butter, or cheese.

Following his Dresden I report, Sternglass developed additional data purporting to show that similar increases in infant mortality could be attributed to rises in the releases of radioactive wastes from the Nuclear Fuels Service Plant of General Electric Company, near West Valley in Cattaraugus County, New York, twenty-five miles south of Buffalo; from the Indian Point I reactor on the Hudson River in Westchester County, north of New York City; and from the AEC's experimental reactor at the Brookhaven National Laboratory in Suffolk County on Long Island. In each instance, he cited infant-mortality increases which coincided with increases in the releases of radionuclides. Only one nuclear-power plant that Sternglass studied seemed to have no effect on infant mortality in the surrounding area. That was the Shippingport plant of the Duquesne Light Company in Beaver County, northeast of Pittsburgh. Developed as a demonstrator by the AEC, the Shippingport pressurized-water reactor is a relatively small one, with an electrical output of 150 megawatts. It was built by Westinghouse on the design scheme of the nuclear-submarine reactor, with containment and filtering devices to hold radioactive releases to a tiny fraction of those from the other plants that Sternglass studied. There were no sharp increases in radioactive releases from Shippingport, and Sternglass found, as he expected, no changes in infant-death rates in the surrounding area comparable to those at Dresden, Indian Point, or Brookhaven.

As had his fallout-mortality correlation, Sternglass's reactor-mortality correlation received vigorous criticism. In a paper presented at Richland, Washington, in November 1971, two critics assailed his conclusions as entirely unfounded.[19] They were Dr. Andrew P. Hull of the Health-Physics Division, Brookhaven National Laboratory, and Dr. Ferdinand J. Shore, City University of New York. Their paper, entitled "Standards, Statistics, and Sternglass: Guilt by Association," charged that Sternglass had failed to show a "satisfactory mechanism" for the effects that he claimed, as well as to consider other causes of infant mortality. The authors concluded that his primary data "are instances of statistical fluctuations," and that a detailed analysis of many years of data, rather than selected years, would show mortality changes fluctuating enough to give contrary as well as positive results, adding that

> On balance, there simply is no consistent substantial evidence to support his contentions. The alleged risks being borne by society have not been demonstrated. Even if the correlations which Sternglass claims to have discovered could be demonstrated to be statistically significant, he seems to overlook that this does not necessarily imply a causal connection.

A different viewpoint on Sternglass's findings was reported in a paper on "Statistical Studies of the Effect of Low-Level Radiation from Nuclear Reactors on Human Health" by Dr. Morris H. DeGroot, head of the Department of Mathematical Statistics at Carnegie-Mellon University, Pittsburgh.[20]

Reviewing Sternglass's report on Dresden I, DeGroot found "mild evidence of a positive relationship between liquid discharges and infant mortality, superimposed on a general downward trend." It appeared, he said, that the peak liquid discharge in 1966 corresponded to an infant-mortality rate of 15.9 deaths per 1000 live births above the over-all linear trend. He cautioned, however, that "the discharges may simply be surrogates for some other variables which are the actual causative agents."

DeGroot confirmed Sternglass's conclusion that there was no evidence of a positive relationship between discharges from Shippingport and infant mortality. However, liquid discharges from Indian Point corresponded to an infant-mortality rate of 2.03 deaths per 1000 live births above the over-all linear trend in Westchester and Rockland counties. And according to the analysis, the increase in liquid radioactive releases from the Brookhaven reactor corresponded to an increase in the infant mortality rate of 4.5 deaths per 1000 live births, in Suffolk County.

Nevertheless, DeGroot warned:

> [The studies] do not present strong evidence that there is a relationship between the exposure of a population to low-level radiation from nuclear-reactor discharges and the infant-mortality rate in the populations—and they do not present strong evidence that there is no such relation. The simple studies carried out here and the inconclusive results do lead, therefore, to a very strong and important recommendation. A large-scale statistical study is urgently needed to aid in resolving this vital issue.

Sternglass's fallout-mortality correlation also received some support from another independent study in 1971. In an examination of "Low Level Radiation and U.S. Mortality," three Carnegie-Mellon University researchers in the Graduate School of Industrial Administration reported findings that fallout is closely associated with mortality rates. The researchers were Dr. Lester B. Lave, professor of economics; Dr. Samuel Leinhardt, an assistant professor of sociology, and Martin B. Kaye, a graduate student. Studying the period from 1961 to 1967, they found "the effect of radiation levels on the mortality rates were substantial" in spite of the relatively poor quality of radiation data, and that cesium-137 is more closely connected with the total mortality rate while strontium-90 is much more closely associated with infant-mortality rates.

For the infant under twenty-eight days, a 10-per-cent increase in strontium-90 is estimated to produce an increase in the mortality rate from a maximum of 0.72 per cent to a minimum of 0.35 per cent, whereas an increase of 10 per cent in cesium-137 is estimated to give rise to a smaller increase—from 0.32 per cent to 0.11 per cent—in the mortality rate.

"These effects are generally substantial and indicate that increases in the levels of these radionuclides are estimated to carry a substantial health cost," the report stated. The researchers added, however, that their findings might be upset by new analyses based on better radiation data, pointing out that it is essential that better data collection begin immediately.

Among the AEC critics of Sternglass, the only one who offered an alternative explanation for the slowing down of the decline of the infant-mortality rate in America was Dr. Tamplin of the Lawrence Radiation Laboratory. In a seminar he prepared for that laboratory in the summer of 1969, a condensed version of which was published in the *Bulletin of the Atomic Scientists* the following December, Tamplin asked: "Is there a reasonable explanation for this leveling trend aside from fallout radiation?" There was indeed, he said. Asserting that Sternglass's data did not support his conclusions, Tamplin said that the major factors influencing infant and fetal mortality over the past fifty years have been the improvement of socioeconomic conditions and the introduction of antibiotics. The dosages produced by fallout radiation simply were not sufficient to have produced the effects on fetal and infant mortality that Sternglass attributed to them. But, Tamplin added, the purpose of his refutation was not simply to criticize Sternglass. He said:

> Frankly, I feel that his article was a naturally evolving social phenomenon. It represents a counter-irritant to the statements of many individuals who profess that exposure to low dosages of radiation is harmless. Contrary to such unsupported statements, the Sternglass article can serve a useful social function. It should cause us to take a serious look at the potential effects of radiation from whatever

source and cause us to focus on the serious problems that represent the basis for the fetal- and infant-mortality statistics in this country.

Arthur Tamplin, one of the AEC's most highly respected investigators into the effects of nuclear radiation on man, had heard the alarm bell which earlier had tolled for Ernest Sternglass.

4
The Ghost of Galileo

Ironically, it was the critique of Ernest Sternglass which Arthur Tamplin developed for the AEC that brought him into conflict with AEC headquarters. The headquarters hierarchy took exception to Tamplin's qualification that the Sternglass correlation served a useful purpose as an exaggerated counterclaim to an equally exaggerated claim that low-level radiation was harmless. That exaggerated claim was AEC dogma. The whole structure of the agency's radiation-protection standards rested on it. And upon those standards were based the design criteria of an entire generation of light-water nuclear reactors in service or about to enter service in the 1970s.

In addition to defending the *raison d'être* of the Sternglass correlation, Tamplin made another and even more serious deviation from the official line followed at the AEC's Germantown headquarters in Maryland. He suggested that fallout radiation did, indeed, kill infants and fetuses—but not as many as the Sternglass correlation had surmised. Moreover, Tamplin continued heretically, "There is strong justification for assuming that the Federal Radiation Council dosage estimates [of radiation from fallout] are low by at least a factor of two . . ." On the basis of his own calculations, Tamplin estimated that the radiation dosage from bomb-test fallout for the years 1953–1957 and 1959–1962 increased the fetal and infant death rate between 3 per cent and 0.3 per cent. These values, he said, were "far below the twofold increase that Sternglass suggests."

In a letter dated September 10, 1969, Dr. John R. Totter,

director of the Division of Biology and Medicine of the AEC, scolded Tamplin as follows, "I cannot agree with your introductory remarks that Sternglass's poor science has served a good purpose and can only assume that it is not your intention to offer scientifically unsound estimates in the hope of serving some meritorious purpose." It was not Tamplin's purpose that he was concerned about, Totter continued, "but the scientific validity of your mortality estimates. I consider your new approach to these to be on even weaker scientific grounds than those presented in your original seminar notes and criticized in my earlier letter."

The last sentence indicates what Tamplin was going through in his effort to produce an honest and unbiased critique of Sternglass under the eye of headquarters. Headquarters was "very disturbed," he related later, that as part of his criticism of the Sternglass manuscript he had estimated the possible effects of radiation from fallout on infant- and fetal-mortality.[1] Tamplin said he had agreed to publish a condensation of his seminar in the *Bulletin of the Atomic Scientists* because:

> It subsequently developed that the major criticisms of the Sternglass manuscript came from the AEC and there the general approach was to indicate that there was no possibility whatsoever that the fallout radiation could have caused any harm to mankind. In other words, Sternglass's somewhat exaggerated claims were being contradicted with even more exaggerated claims by the AEC.[2]

Tamplin discussed the matter not only with Totter but also with Dr. Spofford English, assistant general manager of the AEC, and Dr. Storer, at the Oak Ridge National Laboratory. They wanted him to publish the criticism of Sternglass without making any independent estimate of the possible fetal and infant deaths. By omission, they wanted Tamplin "to support their own incredible position," he related.

This position was summarized in Totter's earlier letter to Tamplin, dated August 21, 1969, in which the division director advised:

In our telephone call, you expressed your feeling that it would be inappropriate to offer a rebuttal of Dr. Sternglass's contention without at the same time offering your calculations of fetal-mortality rates as a part of this rebuttal. We disagree. We feel that these are two separate items and if one is going to propose fetal-mortality risk numbers then it would seem appropriate that such calculations be submitted to a refereed journal, i.e., *Health Physics,* where other scientists would have an opportunity to comment on your lines of reasoning.

Tamplin asserted: "What they wanted me to do was to go ahead and publish my criticism of Sternglass's work in the *Bulletin of the Atomic Scientists,* which is more or less a public journal, but print my own estimates separately in a technical journal that would be read only by professionals in the field." His conclusion was that the request was "phony," and he refused it.

With his longish gray hair and beard, Tamplin bears a cameo resemblance to those seventeenth-century savants who defied the inquisitorial establishment of their time to spread the word that the earth moves around the sun. In several respects, Tamplin's experiences with the Atomic Industrial Establishment of his day have been similar to theirs. So have the experiences of his colleague and former chief at the Livermore laboratory, John Gofman. The AEC's pressure on the two of them to abjure their beliefs about the hazards of low-level radiation have resulted in making their heresy a *cause célèbre.*

At the outbreak of the renewed radiation controversy in 1969, Tamplin was a research scientist in the Biomedical Research Division of the Livermore branch of the Lawrence Radiation Laboratory and Gofman was an associate director. Livermore is a subdued, small town, southeast of San Francisco and Berkeley, lying in a dusty region of craterlike valleys and low, brown hills, partly covered by garish billboards and pale vegetation. In the winter, it presents a lunarlike landscape, modified by an atmosphere, the AEC weapons shop and laboratory, and a four-lane highway.

In 1963, the AEC asked Gofman to set up a biomedical program to study the effects on man of radiation released in the biosphere from the agency's weapons program and peaceful nuclear operations. Gofman did so. Tamplin and a dedicated staff developed extensive and detailed studies on the effects of radiation on man. And these presently convinced them that the release of radiation from bomb or nuclear-explosive tests or from nuclear reactors was barbaric—it was killing fetuses and infants and was increasing the incidence of cancer in the entire population. In attaining these findings, Gofman and Tamplin were doing the job they had been assigned to do by the AEC. The trouble was—they were doing it too well. Their conclusions were going to rock the entire Atomic Industrial Establishment and create a considerable amount of anxiety among officials of the AEC who were dedicated to the development of radiation-emitting nuclear technologies— and who believed that a little radiation was certainly harmless.

The Gofman-Tamplin Manifesto

Gofman and Tamplin formally challenged the safety of the AEC's radiation-protection standards on the evening of October 29, 1969, in a paper delivered at the Nuclear Science Symposium of the Institute of Electrical and Electronic Engineers in San Francisco. They demanded a tenfold reduction in the AEC's maximum permissible-radiation dose to the general population, which was based on the Radiation Protection Guides of the Federal Radiation Council. The demand challenged the authority of the AEC and the wisdom of the hierarchy of scientific organizations that had concocted the radiation standards. Since the light-water-reactor industry in the United States was designing its equipment to meet the standards, the Gofman-Tamplin manifesto threatened the economics of the industry by calling for better radiation containment—especially in the boiling-water reactors, which spewed more radioactive waste products into the biosphere than the pressurized-water machines. The challenge to the scientific organizations was a serious one. If they could not be trusted to protect humanity, who could? These organizations were, in addition to the FRC, the National Council on Radiation Protection and

Measurements and the International Commission on Radiological Protection.

The Livermore scientists warned that if the level of radioactive emissions allowed by the AEC standards under the Code of Federal Regulations was reached by the nuclear industry, it would cause 16,000 cancer cases in the United States a year, in addition to those arising spontaneously in the population. In view of the rapid proliferation of nuclear-power plants, it was only a matter of time, they said, until emissions reached the permissible level. Instead of the recommended maximum population exposure of 170 millirads a year, the exposure limit should be cut to 17 millirads a year. And:

> . . . a hard look at what data do exist leads us to have grave concern over a burgeoning program for the use of nuclear power for electricity and for other purposes with an allowable dose to the population at large of 0.170 rads [170 millirads] of total body exposure to ionizing radiation per year. A valid, scientific justification for this "allowable" dose has never been presented, other than the general indication that the risk to the population so exposed is believed to be small, compared with the benefits to be derived from the orderly development of atomic energy for peaceful purposes.

The basis on which the Livermore scientists estimated additional cancer cases from radiation was the proration of cancer death rates from high levels of radiation, where the rates were known, to low levels, at which the rate could only be surmised in a large population. They contended that all forms of cancer are increased by ionizing radiation and that the increase could be measured in two ways: in terms of the dose required to double the natural incidence of cancer, or as an increase in incidence per rad of exposure.

Further, they said, young persons are more susceptible to cancer from radiation than adults, and the fetus is highly susceptible, as Sternglass maintained. For adults, the dose required to double the natural incidence of cancer is 100 rads—the number called the "doubling dose." If the entire population of the United States was exposed to 100 rads, the number of cancer cases would double at the rate of a 1-per-cent increase per rad. For children, the doubling dose is less, ranging from 5 to 100 rads. For the fetus, the doubling dose is 4 to 6 rads.

If everyone received the "permissible" dose of 0.170 rads each year from birth, each person would have been exposed to 5 rads above natural background radiation by age thirty. This would cause a 5-per-cent increase in the incidence of all forms of cancer a year, at the calculated rate of 1-per-cent increase per rad. In a population of 200 million persons, one-half or 100 million may be thirty years old or older. In this group, the latency period for cancer to develop would be over.

Since the natural or spontaneous incidence of cancer is 280 cases per 100,000 persons, an additional 5 per cent would increase the number by 14 cases per 100,000, or 14,000 cases in the total population over the age of thirty. Add to this another 2000 cases developing in the age group under thirty and the total increase would be 16,000 cases a year from the allowable radiation.

Instead of a 1-per-cent increase in cancer per additional rad of exposure, however, suppose the increase was 2 per cent per additional rad. Gofman and Tamplin believed that this figure probably was closer to reality.

In that case, the AEC standards would result in an extra 32,000 cancer deaths a year.

A Crucial Problem

The two scientists repeated their demand for a tenfold reduction in maximum permissible-radiation exposure of the population at a public hearing on November 18, 1969, before the Senate Public Works Subcommittee on Air and Water Pollution, headed by Senator Edmund S. Muskie of Maine. They told the subcommittee:

> We wish to apprise you that in our opinion, the most crucial pressing problem facing everyone concerned with any and all burgeoning atomic energy activities is to secure the earliest possible revision downward by at least a factor of tenfold of the allowable radiation dosage to the population from peaceful, atomic energy activities.

The Gofman-Tamplin analysis of the hazards of radiation standards reflected a number of studies, including the reports of the Atomic Bomb Casualty Commission on the fate of the Japanese survivors of the bombings of Hiroshima and Nagasaki and the correlations between pelvic X-rays of pregnant women and the incidence of leukemia in their children after birth, found by Stewart in England and MacMahon in the United States.

The Atomic Bomb Casualty Commission (ABCC) found that the leukemia risk was one to two cases per million persons where each person received 1 rad. Since the natural incidence of leukemia is sixty cases per million, Gofman and Tamplin calculated that exposure to 1 rad increases leukemia in a population from 1.6 to 3.3 per cent and the doubling dose is 30 or 60 rads.

The ABCC studies indicated a doubling dose of 100 rads for thyroid and lung cancers, or a 1-per-cent increase in the rates per rad of population exposure. Studies of cancers in uranium miners had indicated a doubling dose of 250 to 500 rads for lung cancer, but the Livermore scientists felt these data were ambiguous. Correcting for latency (the time it takes cancer to develop), the doubling dose suggested by the miners' data could be reduced to a range of 125 to 150 rads. Averaging this with the Japanese data, Gofman and Tamplin obtained 175 rads as the lung-cancer doubling dose. One rad of exposure would thus result in an increase of 0.6 per cent in the annual incidence of lung cancer in the whole population.

The breast-cancer doubling dose was calculated as a 1-per-cent increase in incidence per rad, with a doubling dose at 100 rads. They set the doubling dose for childhood cancer at 4 to 6 rads, basing the calculation on the Stewart and MacMahon surveys of X-ray effects on fetuses.

Gofman and Tamplin proposed several "laws" of cancer induction in humans, derived from their own studies. The first law stated that all forms of cancer can be increased by ionizing radiation, and the correct way to describe the increase is either in terms of the dose required to double the spontaneous rate of each cancer or to increase the rate per rad of exposure. The second law held that all forms of cancer show closely similar doubling doses and closely similar increases in incidence per rad. Their third law stated that less radiation is required to increase the incidence rate by a specified fraction in youthful subjects than in adults.

The possibility that an exposure of the total population to 170 millirads of radiation a year over and above natural background radiation would result in 32,000 cancer cases a year over and above the "natural" incidence struck the Livermore scientists as "a rather high price to consider as being compatible with the benefits of atomic energy." Moreover, they told the Muskie subcommittee, there was also the burden of misery and death from genetic disorders in future generations.

The scientists said they were not against the nuclear generation of electricity:

> We have great confidence that our engineers have the talent to design reactors, reprocessing plants for spent nuclear fuel, transport systems, and waste storage facilities in such a manner that any release of radioactivity that might conceivably expose humans be kept so low as to preclude harm.

Subsequent events demonstrated, as I will relate presently, that the economics of the power industry were more influential than engineering know-how in determining what fraction of the AEC's radiation limit would be emitted into the biosphere by nuclear-power plants. The light-water reactors of the 1960s and early 1970s had been designed under the ceiling of 170-millirads peak emissions. Most of them, particularly the pressurized-water reactors built by Westinghouse, emitted only a fraction of the limit. Nevertheless, economics was the real basis of the risk-benefit equation. Radioactive pollution was a function of the cost of producing power and the cost was inversely proportional to public-health risk.

The Backlash

Reaction to the Gofman-Tamplin manifesto was not long in appearing. Dr. Taylor, president of the NCRP, asserted in a letter to Muskie:

> Their material has presented no new data, new ideas, or new information and quotes only some of the facts now known in this area while other essential data are omitted. The same basic information that they have available to them has been studied continuously by large numbers of highly experienced radiation-protection scientists over at least the last two decades . . . They conclude . . . that the standards should be lowered by a factor of at least ten. This conclusion is obtained on the basis of their adopting predicted potential effects that are thought by most experts to be much higher than any effects which will ever occur.

One wonders what the reaction of Dr. Taylor and other defenders of the 170-millirad dose limit would have been if they had known in 1970 that the AEC, barely a year later, would pull the rug out from under them: it would reverse itself and declare a one-hundredfold reduction in radiation emissions as a design goal for light-water nuclear reactors. But I am getting ahead of the story.

The AEC headquarters generated its own response to the manifesto on December 17, 1969, in a document that was unique in the annals of federal-agency infighting. It was unique in that it exposed the agency in an effort to put down its own experts so as to defend a status quo that was favorable to outside interests—the power industry. The idea that Gofman and Tamplin simply were drawing new inferences from old data pervades the AEC rebuttal. In its formal Staff Response, the AEC said:

> A recommendation to lower the existing standards would appear appropriate only (1) if data have become available that were not considered by the responsible radiation-protection bodies or (2) if valid new interpretations and conclusions have been established through recognized, scientific channels.

To some observers, including me, the logic of such a response seems medieval. New data *had* become available which had not been considered by the responsible radiation-protection bodies—data such as the 1970 report of Stewart and Kneale. The reference to "valid new interpretations and conclusions" carries the implied charge of heresy. Gofman and Tamplin obviously did not go through "channels" so far as the AEC was concerned. The agency channels which should have been open to them were closed, as I shall show presently. Why did they present their data to the Muskie subcommittee before obtaining a hearing before the Joint Committee on Atomic Energy? The whole scientific apparatus of the AEC was arrayed against them.

Furthermore, it is hard to think of a major advance in scientific thought which was not based on new interpretations of old data. Among many notable examples the heliocentric ideas of Copernicus, Kepler, and Galileo that opened the door to modern earth science and astronomy had as a basis the conception of Aristarchus (circa 217–145 B.C.) of a sun-centered, rather than an earth-centered, system of planets. The geographical ideas of Columbus were rooted also in the notion of antiquity that the earth was a sphere. It is difficult to imagine how these ideas and those of Galen, of William Harvey, and of Louis Pasteur might have fared if their acceptance had depended on the recognized scientific channels of the periods in which they were proposed. Were not the observations of Isaac Newton and James Watt available to their contemporaries? Had not the data upon which Albert Einstein developed the special and general theories of relativity been available to other physicists of his time?

The AEC's Staff Response also attacked the Livermore scientists' concept of the doubling dose as being "without scientific validity." This was the same phrase that the staff response to the Sternglass correlation had employed. It is the same charge which, in various forms and languages, has rung down the centuries to dismiss the challenge of innovation and invention. The spectacle of an agency, mandated by law to protect the public safety, stiff-arming its own investigators and denouncing their warnings that its policies may be dangerous to public health is certainly one of the sorriest displays of indifference to the public safety in the brief but stormy career of the AEC. And in the bureaucratic tradition the agency called upon the scientific establishment to support its radiation position against the heresies of Gofman and Tamplin. There are many points of similarity between the reaction of the AEC to the manifesto and that of the Holy Inquisition to the Copernican treatise *De revolutionibus orbium coelestium* 350 years ago—too many to inspire confidence in the agency as a public-health protector. In making the determination of what was scientifically valid and what was not, the AEC payrollers at Germantown were merely perpetuating a tactic which has been employed to discredit new ideas in science since the Middle Ages.

Beyond the pure cant, the AEC Staff Response cited a 1969 ICRP report that the natural incidence of stomach cancer in the populations of five countries varied from 65 to 606 per million. On that basis, the staff argued, the only reasonable way of predicting cancer incidence is the actual number of cases induced by the exposure under consideration.

Gofman and Tamplin ignored wide differences in the spontaneous incidence rates of different tumors for which they calculated the doubling doses, the staff argued. Moreover, estimates of leukemia induction at low-dose rates do not rest on firm, scientific fact, but upon extrapolation from the data obtained above 100 rads. The ICRP was quoted as suggesting that the best estimate of risk is that 1 rad will produce 10 to 20 cases per million persons, or 2000 to 4000 in a population of 200 million. The staff quoted the ICRP further as explaining: "It must be remembered, however, that the carcinogenic effect has been observed in a range of doses of more than 100 rads and these data alone are insufficient to justify attaching any serious credence to the existence of a linear relationship."

The question of whether there is a linear relationship of cancer to radiation has been one of the major points of dispute between the AEC and its critics. A linear relationship means, in effect, that if 100 rads results in 500 cancers, then 1 rad will result in 5. The linear relationship is a hypothesis, which seems to be indicated by some data, but remains neither proved nor disproved. In rejecting it, the staff rejected a conservative approach to radiation protection that assumes radiation effects on a linear basis.

The Bias of Bias

The Staff Reponse charged that the Gofman-Tamplin analysis of the biological effects of low-level radiation was not only scientifically "wrong," but biased as well, because it did not cite the main body of data for human exposure to radiation.

There are twelve general classes of human exposure to radiation which have been discussed in the world medical literature: radiologists operating X-ray machines; the irradiation of patients suffering from a form of spinal arthritis (ankylosing spondylitis); radium-therapy patients; uranium and flurospar miners; patients treated for hyperthyroidism; radium-dial painters (who moistened the brush by putting it between their lips); women treated for cervical cancer; *in utero* irradiation; thymus irradiation; accidental exposures to radiation; exposures at Hiroshima, Nagasaki, and the Marshall Islands; and patients who received a radioactive drug called thorotrast. The staff report contended that the majority of radiation-associated carcinogenesis from these exposures indicates no effect at low-level doses, compared with high doses, the staff obviously ignoring the studies of Stewart and MacMahon on the high incidence of cancer in children who had been X-rayed *in utero*. The staff response concluded: "The opinions and scientifically questionable derivations of Gofman and Tamplin do not make a case for revision of radiation-protection standards."

This opinion was upheld January 25, 1971, when the NCRP announced it had reviewed the radiation standards and saw no need for a change, except to set a limit on occupational exposure of 0.5 rem for pregnant women. Dr. Taylor told a press conference in Washington, D.C.: "Our review of the current knowledge of biological effects of radiation exposure provides no basis for any drastic reductions in the recommended exposure levels despite the current urgings of a few critics." [3]

Threshold

An underlying dispute between the AEC and critics of its radiation policy is whether there exists a threshold below which radiation is harmless. As I related earlier, the concept of threshold was discarded by both the International Commission and the National Council. The Federal Radiation Council observed this dictum. But the establishment of a maximum permissible dose of 170 millirads a year as a radiation standard for the uses of atomic energy implies a contradiction of the concept of zero threshold.

In principle, the AEC accepts the zero-threshold dictum, but allows boiling-water reactors to release hundreds of thousands of curies into the biosphere, with the assurance that these emissions are absolutely harmless. How can this be rationalized?

The National Council on Radiation Protection and Measurements rationalizes it by the doctrine of "acceptable risk." The doctrine holds that society—or rather a certain number of unlucky individuals in it—may pay a price for the benefits of atomic energy. The unlucky ones may get cancer and die. Perhaps, in this point of view, there lurks the vestige of propitiation to certain gods, a form of human sacrifice to the atom gods.

At the level of 170 millirads a year, the estimates of excess cancer cases range from Dr. Taylor's minimum of 85 a year to the Gofman-Tamplin maximum of 32,000 a year. Dr. Karl Z. Morgan, director of the Health-Physics Division, Oak Ridge National Laboratory, has estimated 4000 to 5000 excess cancer cases a year.[4]

In 1970 Morgan advised the Joint Committee on Atomic Energy that the effects of the maximum permissible dose could be verified only by epidemiological studies. He said:

> Any severe somatic injury, such as leukemia, that might result from exposure of the individual to the permissible doses would be limited

to an exceedingly small fraction of the exposed group. Effects, such as shortening of the life span, which might be expected to occur more frequently, would be very slight and would likely be hidden by normal, biological variations. The permissible doses can, therefore, be expected to produce effects that could be detectable only by statistical methods applied to large groups.

Since no large-scale epidemiological studies have been made to prove or disprove such effects, the public-health consequences of low-level radiation exposure remain hypothetical. Why haven't such studies been made? An answer was supplied by Dr. Jesse L. Steinfeld, Surgeon General in the Department of Health, Education, and Welfare in a letter to Representative William S. Moorhead of Pennsylvania dated March 6, 1970. Steinfeld explained that studies to determine the feasibility of a national program to analyze morbidity and mortality data of thyroid cancer, leukemia, and congenital malformations in relation to radiation exposure had led to the decision that a national program was not indicated. The studies on which this decision was based were not published, but the decision was made at the beginning of a rapid expansion in the uses of atomic energy. Steinfeld said that mortality and morbidity reports by state and local health departments, the National Center for Health Statistics, and the National Cancer Institute "continuously provide data, and consultation is openly available from the offices of these sources for investigators."

The consequence of such a do-nothing policy has been to forestall a large-scale effort by the Department of Health, Education, and Welfare to determine radiation effects on the population and to perpetuate the operating assumption of the AEC that a little radiation is a risk anyone can accept. Epidemiological studies by private investigators like Sternglass are more readily attacked than those by an official body.

Morgan suggested to the Joint Committee that the dictum of zero threshold is analogous to a conclusion that there is no time of the day or night when it is absolutely safe for a person who is blindfolded and wearing ear muffs to cross a highway. If he is hit by a car after midnight, when the flow of traffic is sparse, he can be hurt just as seriously as if he attempted this experiment during rush hour. He said:

> Likewise, if a single, ionizing particle happens to strike a cell in the body, bringing about changes such that the cell survives and becomes precursor of malignancy, the victim can suffer serious consequences, just as much as if the malignancy had resulted from exposure to a trillion ionizing particles.[5]

However, the threshold case has never been proved by clinical evidence and no other form of evidence for or against it is fully accepted by a number of experts within the AEC. One of the chief supporters of the theory that there is a threshold below which radiation is harmless was Dr. Robley D. Evans of the Massachusetts Institute of Technology. His reports have formed the scientific authority for the threshold belief among AEC personnel, a position which is important to the whole Atomic Industrial Establishment and the expanding energy market.

In a study of 496 persons with skeletal burdens of radium or of its isotope mesothorium, Evans reported that when the cumulative dose in rads decreased, the tumor time increased for sarcomas and carcinomas associated with radium and mesothorium deposits. This led him to identify a "practical threshold" of dosage, below which the time it took a tumor to appear exceeded the life span of the individual.

Evans reported that the fraction of persons with malignant tumors from radium or mesothorium radiation was roughly the same for all dosage levels between 1200 and 50,000 rads and was zero below 1200 rads. These individuals had been studied since 1932. They acquired burdens of radium and mesothorium in their bones as dial painters, by swallowing radium-dial paint; as chemists, by inhaling or ingesting radioactive compounds; as patients, receiving intravenous injections of radium chloride or ingesting radium water; or as persons drinking various radioactive nostrums, especially one called Radithor.

Of 5000 persons in this group, 452 were studied while living and the cases of 44 were studied after death. Evans concluded that at 700 rads, malignancies might be assumed to develop over a period of forty-five years; at 620 rads, over a period of fifty years; at 400 rads, seventy years; and 250 rads, a hundred years.[6]

Evans' contention that he had demonstrated a threshold in these studies was attacked by Gofman and Tamplin. The fact that no cancers appeared at low dosages in this group, they said, does not mean that none would have appeared in a larger group. In a small group such as this the number of cancers calculated by a linear extrapolation would be less than one. In this group, therefore, no cases would appear. In a larger group, however, the fraction would resolve itself into whole numbers.[7]

Another Livermore scientist held a different view. Edward Teller, the famed proponent of the development of the hydrogen bomb, had become an energetic advocate of Plowshare, the use of nuclear explosives for engineering purposes. Teller had proposed that nuclear-power plants be built underground as a safeguard against a disastrous accident that might spew radioactive debris over the countryside—a view espoused also by Gofman and Tamplin.[8] But Teller was in basic disagreement with them about the effects of low-level radiation. He opined that a tenfold reduction in the maximum permissible radiation dose "would hardly save any lives," in a paper he sent to Senator Mike Gravel of Alaska, at the end of 1969. "On the other hand," Teller said, "such an action would result in the loss of considerable benefits and would give rise to needless complications." [9] The weakest link in the Gofman-Tamplin argument, he said, was the assumption that the entire population would be exposed to the maximum permissible dose. "It appears to me that danger from radioactivity is very much lower than danger from smog, sulfur dioxide, and water pollution," he concluded.[10] This position supported Plowshare experiments which released varying amounts of radioactivity to the atmosphere.

The AEC's attitude toward any challenge to its radiation-protection standards was defined during public hearings on the environmental effects of producing electric power on October 30, 1969, the day after the Gofman-Tamplin manifesto was delivered in San Francisco. Commissioner Ramey, former executive director of the

Joint Committee who was serving his third five-year term on the Commission, testified:

> As you know, Mr. Chairman, I am not a scientist but I have been in this game for a long time. I went through the whole fallout controversy and other controversies when I was staff director of the Joint Committee. I may have gotten a pretty good ear for detecting phonies and phony arguments.[11]

Referring to a recent report to the AEC by its Advisory Committee for Biology and Medicine, Ramey quoted the ACBM report as follows:

> Recently published reports in both the scientific and the popular press claim that past fallout radiations have been responsible for inducing cancer and for increasing stillbirths and infant mortality. There is further implication that the present standards for peacetime atomic-energy uses are unsafe. It is clear to the ACBM from a review of pertinent scientific information that these claims are erroneous and reflect a biased selection and incorrect statistical treatment of the available data. The ACBM is convinced that the data presently available give an adequate foundation of knowledge for setting public-safety standards for radiation. Moreover, there is good reason to believe that future research will reveal that the present standards are conservative.

The ACBM comments had preceded the Gofman-Tamplin manifesto and referred to the storm of concern raised by the Sternglass correlation. But they reflected a mood of certainty that all was well with the radiation-protection guides and the standards. Critics of the AEC position thought they detected a touch of arrogance in the ACBM conviction that "data presently available give an adequate foundation of knowledge for setting public-safety standards for radiation," inasmuch as the precise mechanism for radiation damage to living cells was not known and the effects of the lowest levels of exposure were in dispute.

Commission Chairman Seaborg remained publicly detached from the radiation controversy. Seaborg, an important figure in the Manhattan Project which had built the atomic bomb, was one of the founding fathers of the atomic age and the nuclear-energy revolution. His chairmanship of the AEC since 1961 had spanned its most successful period. Seaborg's stewardship of the AEC had been fairly trouble-free until 1969, when the Sternglass correlation signaled the gathering environmentalist storm that was blowing up around the proliferation of nuclear-power plants. To Seaborg, who looked upon man's control of atomic energy as an advance as important as man's control of fire, the radiation issue was based on ephemeral fears that would eventually dissipate themselves.

In his testimony before the Joint Committee October 29, 1969, he said:

> Now, since we are eventually going to live in a world that will have to depend on the energy of the atom, we must learn to live with the atom wisely. This means we must recognize, anticipate, and deal with all the environmental aspects and prospects of nuclear energy. I believe we are doing this and doing it well. We who are involved in developing nuclear power to provide for future electricity needs are naturally disturbed by that public resistance which seeks to halt or slow down such development. However, along with our obligation to safeguard the natural environment we also have a responsibility to help supply our people with the power to run a technologically sustained society.[12]

Reprisal

On January 28, 1970, two months after they testified before the Muskie subcommittee, Gofman and Tamplin appeared before the Joint Committee and repeated their demand for a tenfold reduction in the maximum permissible dose. In the meantime, the AEC had begun to react to the heretics in the Inquisitorial manner.

Tamplin had been invited to give a lecture at the annual meeting of the American Association for the Advancement of Science (AAAS) during the Christmas week of 1969 in Boston. Normally, this would have been a routine matter, but after he had appeared before the Muskie subcommittee, Tamplin's relationship with his superiors at the laboratory deteriorated rapidly. Its director Michael S. May ordered Tamplin to submit his writings and speeches in advance for AEC perusal—on orders from headquarters. The purpose was not censorship, May had assured Tamplin, but merely to give the agency a chance to answer any criticism that a paper by Tamplin might heap upon it. That seemed fair enough to Tamplin. He submitted his AAAS paper for inspection on that basis.

However, when Tamplin got the paper back, he found that it had been censored drastically. He showed it to Gofman, who remarked drily that about all that remained intact were the prepositions and the punctuation. After protesting to May, Tamplin reported:

> Dr. May said I shouldn't feel bad about the censorship—after all, he wasn't firing me as the AEC had told him to do. I said it was the end of the LRL [Lawrence Radiation Laboratory] as a scientific laboratory because censorship was incompatible with science. Gofman's view of the matter was more emphatic than mine. He told May that the censorship made the laboratory look like a scientific whorehouse.

May denied that he had made any statement about censorship, and the AEC denied it ever instructed May to fire Tamplin or Gofman, who had been collaborating with him. The AEC then advised Tamplin that if he wanted to deliver an uncensored version of the paper to the AAAS, he could do so on his own, but without travel reimbursement by the government. Dr. Roger E. Batzel, who had succeeded Gofman as a deputy director of the laboratory and as Tamplin's superior upon Gofman's resignation in September 1969, had dealt with Tamplin on this question. He told Tamplin that the laboratory would sponsor him if he wished to present a scientific paper on the results of his own research, but that if Tamplin wished to present his personal opinion on nuclear power, "he was free to do so as an individual using nonlaboratory resources." [13] In other words, if Tamplin wanted to attack the AEC's reactor program in Boston, he had to pay his own way.

Gofman threatened to take up the issue of AEC censorship with the AAAS. Negotiations ensued and a compromise was reached. Tamplin agreed to delete from his talk a section on nuclear-reactor accidents and a proposal for a five-year moratorium on building atomic-power plants and conducting Plowshare experiments. The laboratory then agreed to sponsor him at the Boston meeting.

It was a Pyrrhic victory for the AEC, for the skirmish, which was well publicized, simply confirmed the censorship charges. Gofman and Tamplin subsequently presented their proposal for the five-year moratorium to the Pennsylvania Senate Select Committee on Nuclear Generation of Electricity August 20, 1970.

The attempts of the laboratory to disclaim such presentations by Tamplin, or by Gofman and Tamplin jointly, have raised a number of questions about the public responsibilities of a government laboratory. Was it being operated in the public interest or in the interest of a self-contained Establishment? How can the public interest be advanced by censoring dissent?

Two weeks after Tamplin delivered his paper in Boston, seven members of his staff were removed from his supervision and transferred to other duties. The group was developing "A Handbook for Estimating the Maximum Internal Dosage to Man from the Fallout of Nuclear Devices." The handbook, too, was taken out of Tamplin's hands. The preface which he had already written for it was revised. The following June, three more members of his staff were transferred to other duties, leaving him with only one co-worker, Donald P. Geesaman, an expert on the radiological hazards of plutonium.

And Then There Was One

When reports of the breaking up of Tamplin's staff were printed in local and San Francisco newspapers, the AEC quickly issued a statement explaining that no one had been fired. People had simply been "reassigned to higher-priority" projects.

Batzel explained that the seven employees who had been working on the radiation dosage handbook were transferred away from Tamplin as a unit and would continue their work. All the laboratory had done was to take away this key piece of work from a scientist whose outlook on the hazards of radiation differed from those of his superiors. Batzel was quoted as saying that Tamplin had agreed to the transfer, but Tamplin said his consent was never sought.[14]

Of the three additional staff people taken away from Tamplin, two were studying immune response to radiation. They were shifted to a radioecology project which Batzel said had higher priority. Finally, a woman information specialist on the staff was transferred too.

A report of the troubles of Gofman and Tamplin was published by *The Washington Post* on July 5, 1970. It alerted consumer advocate Ralph Nader to the plight of two struggling fellow critics. Nader had been aiding an investigation of the AEC by a group of law students from the University of Texas and was well aware of the radiation issue, along with its industrial overtones.

In a letter to Senator Edmund S. Muskie, Nader raised the issue of persecution of critics by the AEC:

> As far as can be determined, the two scientists . . . have been accused of no wrongdoing, no violation of the AEC regulations and no scientific dishonesty. Actually, the available indications are that Gofman and Tamplin have been accused of heresy by an agency so committed to the promotion of atomic energy that it has insisted that radiation risks be treated more as articles of faith to be intoned than as propositions to be examined continually.[15]

Moving to the heart of the issue with his customary incisiveness, Nader added:

"The Gofman-Tamplin case is not just a technical dispute. It is fraught with serious questions about the role of citizen knowledge and participation in technologies which provide him with both benefits and risks."

Representative Holifield, then chairman of the Joint Committee, asked the AEC for an explanation of the Gofman-Tamplin complaints. The agency responded with a voluminous report dated July 21, 1970. The document is a case study of how a bureaucracy rationalizes its treatment of dissenters on its payroll. It explained that the transfer of Tamplin's staff to other supervisors was merely a result of a change in priorities, brought about by a cutback in the Plowshare program to develop nuclear explosives for excavation and for oil and gas stimulation deep underground. It said:

> It is a fact that over a six-month period, Dr. Tamplin's staff has been reduced from eleven to one. The question is whether or not these actions constituted reprisal for criticism or were justifiable for budgetary or program reasons. Other than the allegations themselves, we have found no indication that these actions were motivated as reprisals for criticism; on the other hand . . . a finding can be supported that the actions taken were for reasons related to budgetary reductions, allocations of resources to programs of higher priority, and a judgment of relative scientific productivity.

The AEC report commented that the reassignment of personnel working on the radiation dosage handbook was a result of this reallocation. Later, seeming to reverse its field, the report stated: "The Laboratory and the AEC attached considerable significance to the handbook and believed it warranted more attention and supervision than Dr. Tamplin would be able to provide *in view of his shift of interest* (italics added) to low-dose effects and the adequacy of radiation standards."

Thus, two conflicting reasons are advanced for the reassignment of the handbook staff: first, that with the cutback in Plowshare the handbook's priority was reduced, and second, that the handbook was so significant a project that it had to be taken away from Tamplin when he became concerned with low-level radiation effects.

A third view of the matter also was offered by the report, which stated: "In discussions with the Lawrence Radiation Laboratory, Livermore, management, Dr. Tamplin acknowledged that this work [the handbook project] was moving along well. He also agreed that Dr. Yook Ng had primary responsibility for the preparation of the handbook and was now ready to carry on this work independently."

Here, indeed, is a rare view of bureaucratic explanation involving several mutually exclusive accounts. The reader is free to select the account which appeals to him. On Tamplin's preoccupation with the radiation standards as a reason for terminating his immunology study, the report elaborated as follows:

> The further reduction in Dr. Tamplin's groups made in late June, 1970, involving the transfer of two staff members reflects the Laboratory's judgment that an immunological study which Dr. Tamplin had been permitted to initiate appeared to lack the requisite supervision, showed little progress, and was of low priority in comparison with other program needs.

When these people were transferred, only two were left.

> In the Laboratory's judgment the size of Dr. Tamplin's remaining unit did not warrant a full-time clerk-secretary. Accordingly, the clerk-secretary was transferred to a comparable position in another section of the Laboratory at the same salary. She was in no sense "fired" as some of the news articles alleged.

Tamplin said the "clerk-secretary" was actually an information specialist who had been invaluable to projects he was working on.

And then there was one—Geesaman.

Operation Truth Squad

The AEC report did not deny the allegations of Gofman and Tamplin that they must carry their message to the public at the private expense of citizens' groups, whereas scientists purportedly representing the AEC travel at the government's expense. That complaint, the report acknowledged, "appeals to a sense of fairness, but it is only part of the story. It would be hard to find a circumstance of employment with the latitude and resources which responsible scientists are provided at LRL-L."

In witness whereof, the laboratory averred that it had duplicated 27,000 copies of twenty-one tracts that Gofman and Tamplin wrote, most of them attacking the AEC radiation standards or citing the hazards of low-level radiation.

Whether the duplication of these materials was an oversight or whether it was consciously permitted may be debatable. The fact is, however, that Gofman and Tamplin had full use of the laboratory's duplicating machine. No one posted a guard or otherwise tried to interfere with their dissemination of their views.

There were other official interventions, however, according to Gofman and Tamplin. May had tried to dissuade them from appearing before the Muskie subcommittee, and the laboratory had docked Tamplin three days' pay when he addressed the Center for Democratic Institutions at Santa Barbara and two days' pay when he spoke at an American Cancer Society seminar at San Antonio, Texas, in March 1970.

Senator Muskie inquired of the AEC if there had been any pressure to prevent the two scientists from testifying before his subcommittee, which was holding hearings on the effects of underground nuclear explosions by the AEC in a series of Plowshare experiments. The reponse, as detailed by the staff report, stated: "Drs. Gofman and Tamplin were told by Dr. May that testimony before the Muskie Committee would have to be clearly presented since this highly technical area might be new to some members of the committee. At no time have they been asked to decline to appear before this committee of Congress." The staff report denied also that Tamplin had been "docked." Tamplin simply had been charged vacation time—one day for the Santa Barbara trip and two days for the San Antonio appearance. "The laboratory does not support staff members for trips to . . . meetings which they attend as individuals promoting their personal points of view," it said, explaining that only attendance at scientific meetings was subsidized.

Feeling the groundswell of rising public concern about atomic power, the AEC had begun a counteroffensive in the spring of 1969, to publicize the benefits of atomic energy. The agency had sent its staff spokesmen all over the country to counter nuclear-power opposition by citizens' groups which were "butting in" at reactor-licensing hearings. These were not scientific meetings, either, but no staff spokesman was docked and all traveled at government expense.

There was a suspicion among some members of the AEC and the Joint Committee that all the spreading public concern was the work of agitators, probably professional ones. In his remarks before the Joint Committee on October 30, 1969, Ramey related that he had noted a convergence of "certain factors" at citizen-protest meetings he had attended in Minnesota and Vermont. "One of these factors is that there are some professional 'stirrer-uppers' involved in each one of the meetings," he said.[16]

Before it took up the cudgel against its own dissenting scientists and others who challenged its radiation-protection dogma, the AEC had gone to extraordinary lengths to sell atomic-energy programs to the people for what it assured them with missionary zeal was their own good. The agency had operated a public-information program for more than twenty years. In many respects, the program set a high standard for clear and well-written exposition and a hard-working public-relations staff at its Germantown headquarters became an important resource for the mass media.

The agency maintained a film library of 11,000 prints. It had issued about fifty annual and semiannual reports, all printed in the austere style of the government printing office but composed in the confident, optimistic prose of the annual investor-utility report to the stockholders. From Chairman Seaborg on down the line, agency officials had made hundreds of speeches, held scores of news conferences. In spite of these efforts, it appeared that the message just wasn't getting through. To their dismay, the commissioners noted that the image of their agency was deteriorating. It was becoming a "spoiler" of the environment, a "tool" of the utilities. Citizens' groups began to regard the AEC with distrust, possibly because the agency tended to regard them as its foes.

Hence in March 1969, the agency began to step up its information program. Howard C. Brown, Jr., assistant general manager, declared the policy would be to "confront the public with the facts." Members of the agency and the staff moved out into the "field" to talk to people. They attended thirty-nine public meetings on the environment, delivered twenty-two speeches, and appeared at ten congressional hearings, submitting more than 300 pages of testimony in the space of eighteen months, Brown related in a speech delivered on February 11, 1970, at the Atomic Industrial

Forum's Topical Conference on Nuclear Public Information at Los Angeles. The staff prepared sixty-six articles on environment-related matters, and more than 140,000 copies of an AEC booklet entitled *Nuclear Power and the Environment* were distributed. The booklet said that nuclear power was the clean way to generate energy—it did not discuss the pros and cons of the radiation issue.

Brown related that the headquarters-information effort was doubled from thirty-five to seventy "man years." The staff traveled the length and breadth of the land, preaching that old-time, atomic gospel. There was a note of patient resignation in Brown's account, for, while the bearers of the true tidings confronted the "antinuclears" at every turn, the unbelievers were stubborn and unconvinced.

One of the early stops of the "forensic trail" was Brick Township, New Jersey, on June 25, 1969—a place thought to be much in need of atomic energy. By that time, the Sternglass correlation had hit the mass media and the alarm was spreading.

Brick Township, said Brown, "was really our initial public confrontration after the Commission decided to take our case directly to the public." The AEC had been asked to present its case by the League of Women Voters, but "we found the meeting was being organized by two well-known, arch antinuclears." Brown continued: "We presented a scholarly and factual case but we weren't nearly as entertaining as the opposition." The magazine *Nucleonics Week* characterized the bout as "a technical knockout—of us!" Brown admitted.

The AEC traveling truth squad believed it was its performance, not its case, that was deficient in public persuasion, Brown said, and "we spent the summer [of 1969] studying our mistakes and preparing for the next encounter, which turned out to be Burlington, Vermont, on September 11, 1969."

Alas, Burlington also was a disappointment. The art of presenting the AEC case still eluded the official road show. Besides, Brown explained, somebody switched the signals at Burlington. Instead of the lengthy, scholarly discussion which the AEC speakers were prepared for, everyone was limited to ten minutes. Questions from the audience seemed hostile. Brown went on:

> The next two rounds were at Brattleboro and Bennington. . . . Now, contrary to some of the news reports, the situation at Brattleboro was that we had to pull our punches to avoid embarrassing some of the opposing panel members. Bennington turned out to be a high-school-style debate. Our opponents skipped from topic to topic so fast one couldn't remember what they said, except that it was bad. Rebuttal was difficult because the facts were vaporous and elusive.

At Minneapolis, Brown said, the case for nuclear power "received a tremendous assist" from Commissioner Ramey, California's Representative Hosmer of the Joint Committee, and several employees of the agency "who discredited extravagant, doomsday forecasts." He cited the report of *Nucleonics Week* that the Minneapolis junket "was a winner, not a big winner, but rather a by-a-nose winner."

So it was that the agency, which denied its critics Gofman and Tamplin travel funds and time off to present dissenting views, financed a promotional tour by staff members of more than 200,000 miles to refute dissent.

Brown said he was optimistic, now, that "the energy story is getting through." Also, the AEC's allies were shaping up:

> . . . the nuclear industry is much better organized than it was a year ago. Look at what's happened in the past year. Westinghouse's new course of instruction on environmental matters; the General Electric seminars; this conference and other Atomic Industrial Forum programs. Note what the utilities are doing to organize their resources to deal with the problem. We are also making an "honest gal" out of the opposition, because the word is out—that wherever misstatements are made, we'll counter with the facts. By "we" I mean the Commission and the industry.

But all of these efforts to soothe the populace, to repress dissent, and to neutralize environmentalist opposition to the proliferation of nuclear-power plants had no effect. Citizen opposition spread and became more determined during 1970, and the AEC began to reconsider the radiation-protection standards that its agents had sought so diligently to defend. Under the public pressure and criticism which it so disdained, the agency was being compelled to yield, but its surrender the following year would be, to say the most, limited and conditional.

5
The Intervenors

The proliferation of nuclear-fission power plants in the second half of the 1960s was accompanied by a swelling chorus of grass-roots concern about radiation and thermal pollution. Much of it was aroused by the environmentalist movement and the new awareness it generated of the threat of all forms of pollution to the survival of the increasing human population. Radiation had long been established in public consciousness as a dreadful poison which not only could deal death and disability to the living but alter the humanity and viability of generations yet unborn.

In such states as Pennsylvania, New York, Illinois, Michigan, California, and Minnesota, the domes of nuclear-reactor containment structures appeared—symbols of the growing economic power of the Atomic Industrial Establishment and the radiation threat it presented in the American transition from a hydrocarbon to a nuclear-fuel economy. Old distrusts of electric-power utilities were revived by critics of the spread of nuclear power. Added to these was a lack of confidence in the credibility and regulatory independence of the Atomic Energy Commission in dealing with the private power industry.

As the official promotion agency of atomic power, the AEC had acquired the image of an ally and partner of private interests, thus becoming an adversary of public demand for tighter radiological emission standards. The role had developed over the years—out of AEC concessions in licensing reactors as "research" facilities, thus enabling powerful utility groups to escape a prelicensing antitrust review of their power-development plans for years; out of the agency's efforts to suppress or discourage scientist critics of its radiation-protection policies, as though the standards were Holy Writ; out of the agency's conflict of interest as both promoter and regulator of a technology whose environmental effects were in dispute. Instead of a judicious attitude encouraging the fullest possible public debate of a controversy affecting the health of 200 million Americans and of their descendants, the AEC adopted a narrow, bureaucratic, and vindictive policy toward critics of its radiation position and drafted legislation to restrict the intervention of citizens in nuclear-power-plant licensing hearings. In the developing environmental drama, in which concerned citizens wore the white hats, the AEC was cast as one of the spoilers.

The response of worried citizens to the AEC-private-utility partnership was a tactic of intervening in AEC construction and operating-license hearings for new nuclear-power plants. The intervenors could be anybody: groups ranging in size and influence from neighborhood civic clubs and *ad hoc* organizations to powerful state, regional, and national bodies of sportsmen, wildlife protectionists, and conservationists.

While intervenors could not, in most instances, prevent the construction or operation of a new plant, they could delay it, thus wringing environmental concessions from utilities—concessions which the AEC did not require. Postponed indefinitely as a result of citizen protest against the thermal pollution of Cayuga Lake was the proposed Bell Nuclear Power Station of the New York State Electric and Gas Corporation at Milliken Station, on the east shore of the lake.

In a referendum, the citizens of Eugene, Oregon, voted to cut off funds to the Eugene Water and Electric Board for the construction of a nuclear-power plant until its environmental effects could be determined. Plant construction was postponed at least until 1974. Citizen protest in California persuaded the Pacific Gas and Electric Company to withdraw its 1964 application to build a nuclear generating plant on Bodega Head, a promontory adjacent to the San Andreas rift system about fifty miles north of San Francisco.

Grass-Roots Regulation

Citizen intervention developed a technique of its own to stall construction and to delay operation until the utilities conformed to some extent to environmental-protection demands to which the AEC and other agencies were indifferent. The utilities discovered quickly that they could not write off the intervenors as cranks. Any attempt to ignore or harass them brought them waves of public support.

Before either the industry or the public was fully aware of it, citizen intervention evolved into a new, grass-roots, regulatory device. Such a device never was anticipated by the AEC, the Joint Committee on Atomic Energy, or the atomic-power industry. But by early 1971, it was forcing power utilities to agree to thermal and radioactive waste standards far below the levels required by the AEC or other public agencies. Thus, intervention became the citizen's weapon against the Establishment.

The effect of this device was to increase the cost of nuclear-plant construction by requiring the installation of pollution-reducing equipment not required under AEC regulations. This, the industry found, affected the cost competitiveness of nuclear plants versus fossil-fuel power plants. Utilities which had decided to go nuclear to avoid increased costs of low-sulfur fuels—to which many were restricted by antipollution regulations of cities and states—found they faced the same kind of environmental taxation if they built nuclear reactors. For while the reactor manufacturers designed plants to conform to AEC requirements, the additional requirements by the intervenors involved extra expense. Utilities agreed to meet the demands of the intervenors only because it was less costly to add the new, antipollution equipment than to delay power production while the issues were thrashed out in the courts.

If public intervention could persuade the power companies to reduce environmental pollution, why did the AEC persist in maintaining a more permissive radiation-emission standard? Was the agency's promotional bias overriding its regulatory responsibility?

There is hardly any question about it. Since the first generation of fission reactors in commercial use was designed to meet that standard, a reduction in the standard would have impact on the entire industry. Besides, from the official AEC point of view, no reduction was necessary because the amount of radiation it permitted was either harmless or, at worst, a small hazard in exchange for a large benefit—the benefit of electricity.

Under this administrative philosophy, dissent was equated with disruption. Dissenters were "crackpots." They were in the AEC-industry view "unscientific, rabble rousers, publicity seekers, alarmists, fanatics." In twenty-five years, the civilian agency to which the atomic scientists of Chicago had struggled so hard after World War II to transfer the control of atomic energy from the Military Establishment had itself become an Establishment—as rigid, as secretive, and as intolerant of criticism as its military predecessor had ever been.

The AEC's radiation-protection standard allowed a maximum of 500 millirads of radiation exposure a year per individual at the plant boundary. Since radiation dosage is cumulative, a large number of nuclear plants emitting radioactivity at this level could raise the average radiation dosage to the whole population to its legal limit—170 millirads a year.

The difference between 500 millirads and 170 millirads should be noted. The first number is the specific maximum dose any person could receive legally in one year from a nuclear reactor at the plant fence. The second number, as I explained earlier, is the average each individual in the entire population of the United States might receive in one year from all nuclear-power and fuel-processing plants, exclusive of any radiation from X-rays, television sets, medical emitters, or cosmic rays. The 170 millirads is one-thirtieth of the 5 rads which both the NCRP and the ICRP say is the maximum an entire population should be allowed to receive over and above natural background and medical radiation in a generation—thirty years. This amount is based on the maximum permissible dose the reproductive organs might receive without undue risk of causing genetic aberrations in offspring. All radiation involves some risk and the AEC view is that the risk must be balanced by some public benefit. Is the risk of 170 millirads a year to the gonads and ovaries balanced by benefits of nuclear power, as compared with coal-, gas-, or oil-power plants and their emissions? The AEC's official position is that it most definitely is more than balanced, but the critics point to the statistical probabilities of increased cancer and infant mortality, in addition to the genetic risk.

For the individual, the mythical being presumed to be lurking continuously about the plant gate, 500 millirads is considered an acceptable risk—that is, acceptable to the AEC. No one has been known to get cancer from such a dosage and no genetic effects have been observed in nuclear-plant workers since the Shippingport plant began operating west of Pittsburgh in 1957. But when the risk is spread over the whole population, it is reduced to 170 millirads. And once the individual lurking about the plant gate leaves and goes off to join the population beyond plant, or "off site," he becomes subject to the 170-millirad standard. It is a strange, often incomprehensible system of public protection, and it is largely fictitious.

Fission reactors in use today do not emit radioactivity anywhere near the 500-millirad-dose limit. They emit only a small fraction of it—most of the time. There are times when the emissions rise —when fuel rods develop leaks, as they inevitably do, or when they are being removed to be shipped to the reprocessing plant.

The reality is that the AEC standard for dose at the fence allows a big margin for error in the design and operation of the reactor—for each plant. None of those operating at this writing is likely to exceed the limit unless there is an accident. But while the 500-millirad limit allows margin for the utility operators and reactor manufacturers, it does not, in the opinion of the critics, give that kind of a break to the public: it does not give the human part of the equation a margin of safety. What margin exists is the ability of nuclear plants to keep their emissions far below the legal fence limit. But that margin is achieved by a *modus operandi* and by engineering; it is not achieved by any emission regulation of the AEC. The agency's permissive regulations betray a rigid, bureaucratic indifference to public concerns, and when civic groups or individual citizens come to realize this, they tend to regard the agency as their adversary.

Both industry and AEC spokesmen have stated publicly that it is not only possible to design nuclear reactors so that their emissions of radioactivity are 1 per cent or less of the plant-fence maximum, but also that the new generation of fast-breeder reactors is being designed in this way.

The conflict between public safety and cost in the radiation emissions of power reactors was identified as early as 1958 in a study of radiation standards by the National Advisory Committee on Radiation. The preliminary draft of its report was submitted to President Eisenhower's science adviser, James R. Killian, Jr., in the fall of 1958 and was openly critical of the AEC in several respects. It said:

> The dual role of the AEC in the promotion and development of atomic energy on the one hand and its regulation of radiation on the other is an interesting one. Generally, the vesting of both promotional and regulating functions in the same agency is unwise and may be expected to present a number of serious difficulties. Foremost among these is the possibility that the agency in its zeal to carry out its promotional activity may lose sight of its responsibilities in operational safety.

The draft went on to say:

> During its lifetime, the Atomic Energy Commission on a number of occasions has indeed been accused of subordinating radiation safety to economic advantage when several of its nuclear-reactor installations have been planned. In the case of a power reactor constructed at Sandusky, Ohio, for example, considerable concern was expressed by a number of individuals and groups over the establishment of such a large reactor in the center of a densely populated area. Whether these criticisms were well founded or not, it is noteworthy that the dual responsibility of the Atomic Energy Commission has proved embarrassing and it may be expected to prove increasingly so in the near future.

Some modifications were made in these statements after the draft was reviewed by the AEC, the Department of Defense, and the National Bureau of Standards.[1] The last section was altered to read: "It is noteworthy that the dual responsibility of the Commission has been the cause of not inconsiderable misunderstanding."

Efforts to reduce the conflict of interest by transferring the AEC's radiation-protection responsibility to the Public Health Service were opposed by the agency and a compromise was reached in the creation of the Federal Radiation Council (FRC) in August 1959 by President Eisenhower, as mentioned in Chapter 3. The Council was mandated to consult qualified experts in radiation physics and biology. In its first set of radiation-protection guidelines, the FRC permitted both Department of Defense and the AEC to set their own radiation standards by stating: ". . . the guides may be exceeded only after the Federal agency having jurisdiction over the matter has carefully considered the reason for doing so in the light of the recommendations of this paper." [2]

Criticism of AEC standards by citizen groups, scientists, and the press persuaded the Nixon Administration to vest the standard-setting function in the new Environmental Protection Agency (EPA). But this transfer was meaningless, since the EPA simply absorbed the AEC's standards and administrative personnel. Nothing had changed except the demise of the Federal Radiation Council. If citizens or scientists were concerned about the pollution allowed by the standards, their only immediate recourse was to intervene in license hearings. Even then, recourse was limited to the radiation question. The Atomic Safety and Licensing Board refused to consider complaints about thermal pollution, contending this environmental problem lay outside the agency's area of control.

On Long Island, a skirmish in the battle of the intervenors against the nuclear Establishment produced a remarkable debate about the nature of the radiation hazard. The conflict arose from the application of the Long Island Lighting Company to construct a $271-million, 820-megawatt boiling-water reactor at Shoreham, on the north shore of Long Island about twenty miles from New Haven, Connecticut.

Intervenors had raised a number of issues, among them the catastrophic effects of a possible aircraft collision with the plant, which was to be built in an area of heavy airport-approach traffic. But the principal issue was the radiation hazard of a plant allowed to emit radioactive pollution under the existing AEC standards.

In the Long Island hearing before the Atomic Safety and Licensing Board, one group of intervenors confronted another. The Lloyd Harbor Study Group, composed of residents of that community, opposed construction of the power plant. The other intervenor, an organization called the Suffolk County Scientists for Cleaner Power and a Safer Environment, favored it. Most of the members of the Suffolk County Scientists were employees of the Brookhaven National Laboratory, the AEC research center at Upton, New York, and the "cleaner power" phrase in the name of their organization referred to nuclear power rather than fossil-fuel power. Consequently, they supported the applicant, the Long Island Lighting Company.

The debate waxed intermittently for months during the winter and spring of 1971 at public hearings on Long Island.

Against the Suffolk County Scientists, the Lloyd Harbor laymen recruited a number of experts to support their contention that the plant would be a radiation hazard. One was James D. Watson, winner of the 1962 Nobel Prize in physiology and medicine with Francis H. C. Crick and M. H. F. Wilkins for discovering the structure of deoxyribonucleic acid (DNA). Watson was director of the Cold Spring Harbor Laboratory on Long Island, where he was studying the connection between viruses and cancer.

He testified that the way in which radiation induces cancer is still fundamentally unknown, but it most likely sometimes activates cancer viruses. In humans, he said, "we must assume very lengthy incubation interludes of the order of twenty to forty years before many radiation-induced cancers are likely to arise." Considerable heterogeneity in human susceptibility to radiation-induced cancer exists, he said, and "we must assume a linear connection between dose and probability of causation." He added that the amount of research now being done on the connection between cancer and radiation is totally inconsistent with proposals for widespread introduction of nuclear plants into highly populated areas.

Watson observed that the 170 millirad standard as the maximum average population dose was not "the great figure" its proponents claimed:

> Slowly and surely our society will be educated to the fact that you are starting from the wrong base lines. As long as you start from that base line, things sound good. But you are starting from the wrong figure and therefore you can trap people into saying it sounds good, but it is not good.

Asked by opposing counsel whether he could suggest an alternative to the use of nuclear power, Watson said:

> I think the current expenditure of the AEC on fusion research is unrealistically low and will lead us to this disastrous breeder program, in which the filth of radioactivity will be [with us] for millenniums, you know. The AEC is not supporting the type of research it should in order to give us a chance of having anything like clean power.

Another witness for the Lloyd Harbor Study Group was the British physician Dr. Alice Stewart, whose study of the correlation between pelvic X-rays of pregnant mothers and leukemia in their offspring was cited earlier as one of mainstays of the argument that low-level dosages of radiation from nuclear bombs and power plants increased cancer incidence in a population. In her testimony she said:

> Since 1956 I have been directing a nationwide survey of children's cancers. It takes us over Scotland, England, and Wales and reckons to collect every child that dies of malignancies and obtain family histories and a large body of controls. It was as a result of this work that we first read, I think it is said first recognized, the low-dose radiation cancer rise in man and we have since gone further and have been able to show . . . absence of the safety threshold. The significant point, I think, of this evidence is that at these very low doses there seems to be this linear relationship that no doubt you have heard a great deal about. Before this evidence became available it was only possible to extrapolate from doses of the order of about

100 or more rads, but this had conclusively shown that in the range of less than 1 rad this rule still holds true.

Of the experts testifying in behalf of the Lloyd Harbor Study Group, the one whom the AEC and the Brookhaven scientists tried hardest to shoot down was Dr. Sternglass. The University of Pittsburgh physicist presented data from which he inferred that anomalous local rises in infant mortality could be attributed to radioactive pollution from nuclear-power plants at Peach Bottom, Pennsylvania; Big Rock Point, Michigan; and Dresden, Illinois; and from the Indian Point and Brookhaven National Laboratory reactors, along with the West Valley fuel-reprocessing plant, in New York.

The allegation that there was a correlation between emissions from the AEC's experimental Brookhaven reactor and a rise in infant mortality in Suffolk County hit the Suffolk County Scientists for Clean Power right in their employment. They mobilized their best data to discredit it.

The Sternglass correlation for Brookhaven ran as follows. From 1953 to 1955, the infant mortality rate in Suffolk County rose from 19.5 to 23.5 deaths per 1000 births. In this period, radioactivity emitted by the Brookhaven reactor in liquid wastes increased from 35.8 microcuries in 1953 to 75 microcuries in 1955. Again, as radioactive emissions rose from 106 to 219.1 microcuries between 1958 and 1961, the infant-death rate in the county jumped from 21.1 to 24.1 deaths per 1000 live births from 1960 to 1961. The infant-death rate then declined to 19 per 1000 in 1964 when radioactivity in liquid wastes dropped to 76.4 millicuries per year. It reached a low of 13.1 per 1000 births in 1969, in parallel with a drop in radioactive emissions from Brookhaven to a level of 16.2 microcuries by 1968.

No such fluctuation of infant-mortality rates was observed in the same years in neighboring Nassau County, in Fairfield County, Connecticut, across Long Island Sound, or in New York City, Sternglass said, but a similar correlation between radioactive releases to the environment and infant mortality could be made in the vicinity of the Indian Point nuclear-power station in Westchester County.

The Suffolk County Scientists had their chance to demolish the Sternglass correlation on May 11, 1971, when the Board convened to resume the construction-license hearing at a motel near Centereach, Long Island. Sternglass was cross-examined at length by Vance L. Sailor, a nuclear physicist at Brookhaven.

> SAILOR: Doctor Sternglass, how do you explain the fact that after this reactor started operating the infant mortality in Suffolk County stayed about steady or actually improved until 1954? Are you suggesting that the emissions from the reactor had a favorable effect on the infant mortality of Suffolk County?"
> STERNGLASS: Hardly. I think the more likely explanation is that the early releases were held to very low values. In 1951, only 21.5 millicuries were released for the whole year, but in 1961 it was 219.1 millicuries or almost a tenfold increase in the amount released. And I suggest that the tenfold increase was responsible, or it is highly likely to be responsible, for the increase in infant mortality in and around the Suffolk County area near Brookhaven.

In the early years, Sternglass said, these reactors show very little release of radioactivity. It is only as corrosion proceeds that the large increases in radioactivity take place in the gaseous and liquid effluents. He continued:

> It is my contention, that the good performance in the early years corresponds exactly to the good performance of the Dresden reactor in the first few years and of many of the other reactors we have studied. . . . The initial results are always hopeful but as time goes on, corrosion takes place, the fission products start leaking out, and the build-up of radioactivity out of the smokestack and into the rivers takes place.

The exchange between the two scientists rapidly escalated in intensity from a cross-examination to an argument, with Sailor challenging the Brookhaven correlation and Sternglass stanchly defending it. In effect, the Brookhaven physicist argued that the rapid growth of Suffolk County during the late 1950s and early 1960s was a factor affecting the rate of infant mortality. Sailor then established (by asking Sternglass if he knew) that 90 per cent of the Suffolk County population live west of Brookhaven and only 10 per cent reside to the east—downwind of the reactor and downstream on the Peconic River, into which the liquid wastes from the reactor were dumped.

Brookhaven scientists were aware of all this. Was Sternglass? At one point, Sailor demanded, somewhat fiercely: "Dr. Sternglass, are you testifying that Brookhaven has not conducted a rigorous monitoring program?"

> STERNGLASS: To the contrary. I believe that Brookhaven is the most measured, the most studied, and most monitored laboratory of any in the United States, with the possible exception of Argonne [the AEC laboratory near Chicago]. But the tragedy has been that during all these years, when these monitorings were carried out, no parallel studies of possible health effects on the population were carried out.
>
> SAILOR: Do you know whether anyone drinks water from the Peconic River? Do you know whether any water from the Peconic River is used for farm-irrigation purposes? Do you know whether anyone fishes in the Peconic River? Do you have any knowledge whether people take foodstuffs from the River?

Sternglass, alas, had not made a study of the Peconic River. But the Lloyd Harbor Study Group was not without its resources. A brief dossier on the Peconic River was obtained during a ten-minute recess, and when the hearing resumed, the Group's attorney, Irving Like of Babylon, arose and said:

> Mr. Chairman. The Lloyd Harbor Study Group would like to request that the Board make a physical examination of the Peconic River area. I have been informed . . . that the Peconic River empties or flows partially at least into Swan Pond, which is used to irrigate a cranberry bog. And the cranberry bog is part of a major industry which ships cranberries all over the United States. Second, I am informed . . . that there is shell fishing at the mouth of the Peconic River which is controlled by bay constables under special permits issued by the Department of Environmental Conservation. Consequently, we would request that the Board make an official view of the environs as being a site-related matter that is relevant to the proceedings in the Shoreham case.

Sailor then asked Sternglass if he was aware that the radioactive concentration of the Peconic River at the point of human habitation closest to the reactor cannot be distinguished from that of natural radiation background?

Sternglass was not. Was he, then, aware that most of the gaseous Brookhaven effluent was in the form of inert argon-41?

"It certainly was," agreed Sternglass. "The yearly emission rate . . . was 4,350,000 curies. . . . This was comparable to the release of about 8.7 million curies of fission-product noble gases. . . . And this exceeds any reported emission of fission-product gases from a power reactor in the United States."

> SAILOR: Now can you tell us how argon-41 gets into the food chain?
> STERNGLASS: I did not say that argon-41 gets into the food chain. Argon-41 is inhaled. And under equilibrium levels of high concentration, such as occurred near Brookhaven, it equilibrates with the bloodstream, gets soaked up, in effect dissolves, in the fatty tissue and thereby builds up and causes an internal dose from beta radiation in excess and, in fact, in addition to the gamma radiation from the outside. The mere fact that argon is a material of low chemical activity does not make it innocuous as an agent to be inhaled freely by pregnant mothers. . . .

Sailor then questioned Sternglass about the work by Stewart and Kneale linking pelvic X-rays of pregnant women and leukemia in their offspring. Was Sternglass not aware of a study (by Jablon and Kato) showing that of 1292 children exposed *in utero* to radiation during the atomic bombing of Hiroshima and Nagasaki, only one death was recorded from a childhood malignancy?

> STERNGLASS: . . . children irradiated *in utero* at Hiroshima and Nagasaki were aborted and died of other causes before they had a chance to die of leukemia.
> SAILOR: One would have expected thirty-seven deaths on the basis of Stewart's and Kneale's figures. One death was observed. Doesn't that strike you as peculiar?
> STERNGLASS: Not at all, if you consider all the terrible circumstances under which these studies were done. Many of the children in order to develop leukemia have to live to be ten years old and many of them did not do so.

So ran the debate on the radiation issue.

Another witness supporting the position of the Lloyd Harbor Study Group, Marvin Kalkstein, a physicist at the State University of New York, Stony Brook, argued for lower radiation limits. He called attention to the Long Island Lighting Company's environmental statement which promised: "It is expected that during the operation of the Shoreham Nuclear Station, the off-site dose to individuals will, on the average, be less than 1 per cent of the requirements of the AEC limits. . . . The radioactive liquid releases will also be less than 1 per cent."

Kalkstein commented:

> This is one of the few instances in their environmental statement where the numbers cited have any real meaning and particularly where they can be grasped and dealt with by technicians and laymen alike. It does not seem unreasonable to ask a power company to deliver on the one technical specification that can be thus readily identified and evaluated. . . . Why not license on this basis?

If it was technically possible to keep radioactive emissions from a nuclear reactor to 1 per cent of the AEC's radiation-protection standard, why should the agency allow more? The AEC standard allowing a dose of 500 millirads per person at the plant boundary was based on the doctrine of "acceptable dose"—a dose that did not appear to harm anyone and that could be considered a reasonable risk in exchange for the benefit of nuclear power. The standard did not relate to the capability of a reactor to produce less; it simply restricted reactor operations that would produce more.

In Minnesota, however, a pollution-conscious state government adopted a set of state radiation-protection standards based largely on the capability of light-water reactor technology to hold radiation emissions to one-hundredth of the AEC standard. Consequently, the Minnesota standards for gaseous and liquid radioactive wastes were at least a hundred times more restrictive than those of the AEC.

Minnesota moved to enforce its standards for the first time in 1968 after the Northern States Power Company received an AEC permit to build a $115-million, 640-megawatt (electrical) boiling-water nuclear-reactor power station at Monticello, Minnesota, 40 miles upstream on the Mississippi River from Minneapolis and St. Paul.

Northern States Power balked at the radiation standards in the state's waste-disposal permits. The company asserted that any attempt to meet them would require changes in radioactive-waste systems, already designed in accordance with the more permissive AEC standard. The company maintained that the higher standards of the State of Minnesota—where it was technically possible to meet them—would force the utility to incur excess costs. This position suggested an inverse proportion between radiation emissions and construction costs, or a direct relationship between economy and radiation pollution of the environment.

The company filed suit on August 26, 1969, in the United States Distric Court at St. Paul, seeking a declaratory judgment that the State of Minnesota lacked authority to compel the company to meet its radiation standards. The company contended that under the Atomic Energy Act of 1954, and later amendments, the federal government pre-empted the setting of radiation standards and the control of radioactivity emitted by nuclear processes.

The issue of federal pre-emption in this field, involving the rights of the states to protect life and property, attracted national attention at the outset. Eight other states and the Southern Governors' Conference filed briefs with the District Court to intervene as *amici curiae*.

In its brief, the State of Illinois said, in part: "Indeed, we are puzzled with the comments of counsel for the plaintiff . . . to the effect that wisdom in the area of radiation pollution virtually singularly reposes in the AEC and that it would be idle to argue that the states can improve upon the AEC standards."

The State of Michigan's brief asserted: "Any decision rendered by the court in this case will have decisive impact upon a state's prerogative to maintain its waters and air from pollution from all sources."

In its complaint Northern States Power maintained that it would have to make "substantial alterations" from the design of the plant, as approved by the AEC, to meet the state standard. The result would be a two-year delay in producing badly needed power. It would have an adverse effect on the reliability of the power system and cause "the expenditure of several millions of dollars in excess construction and operating costs."

The company distributed electric power in Minnesota, North Dakota, and South Dakota, and through a subsidiary in Wisconsin. Unless the Monticello plant was in operation by June 1970, the complaint said, "the reliability of electric service in the area served by the plaintiff will be impaired."

The added expense the company would incur if it were compelled to meet the state standards was detailed in the utility's reply to a series of interrogatories. Three basic waste-system additions to the Monticello plant would be required to enable the company to comply with the state agency's permit. First, the radioactive-gas waste system would have to be enlarged to include recombiner equipment, a charcoal absorption system along with auxiliary equipment and additional controls. These additions would trap most of the radioactive gases and would require enlarging the plant building. Second, the radioactive liquid-waste system would have to be expanded to include a concentrator, an additional demineralizer, substantially more tankage and handling equipment, and associated building additions and alterations. Third, a low-level radioanalysis laboratory would be required and would include the development of presently nonexistent precision equipment, located in a new, separate building to insure a low level of background radiation.

All these systems, the company said, do not exist in their entirety in any plant in the world. Therefore, the company could not predict the amount of increased cost and extra time that would be required for each item. But its best estimate placed construction-cost increases at a total of $9,500,000. The gas-waste-system addition would cost $7 million; the liquid-waste addition, $2 million, and the low-level radiation laboratory, $500,000. Whether instrumentation adequate to verify the low radiation-emission levels that the state demanded could be developed "in the foreseeable future" was "questionable," the utility said.

Moreover, the modifications which the state agency required would delay the start of operations. Lost power production resulting from such a delay could cost the company as much as $20 million in a period of two years.

In addition, the company said, compliance with the state permit would require the removal of leaking fuel elements before scheduled refueling. This would incur excess cost because the fuel that was removed ahead of schedule would not have produced its expected amount of energy. The plant would then have increased fabricating costs, increased conversion costs, and increases in new fuel-shipping costs, reprocessing costs, interest costs, and capital costs for spare fuel, the company asserted. It added:

> Compliance with the Pollution Control Agency [PCA] permit would have adverse effects on the reliability of the Monticello plant and therefore on the Northern States Power Company electric system. The availability of the Monticello plant would be considerably reduced, mainly by frequent and lengthy shutdowns, to meet the PCA limits on radioactive-gas discharges. These gas-waste discharges occasioned by very slight fuel leaks would be allowable under the AEC limits and the plant would continue to run and supply to the NSP power system.

Under the state standard, fuel leakage that discharged radioactivity to the environment would require the plant to shut down until the leaking fuel rods could be replaced. AEC standards permitted fuel leakage up to a point. Thus, the state requirements meant increased operating costs which the company said would amount to $1.25 million a year, in addition to the $9.5 million in extra construction costs and the $20 million in lost power production. Also, the company said, plant personnel would be subject to "significant radiological exposure" in handling larger amounts of radioactive wastes and from more frequent start-ups and shutdowns to search for minor leaks in fuel rods and "attendant protection procedure."

The company concluded: "It is impossible to comply fully with all requirements of the PCA permit regarding radioactive waste discharges, although the plaintiff could operate well within the standards promulgated by the AEC."

In this manner was the *Cost* vs. *Radioactivity* issue laid out. In his opinion of March 17, 1971, Chief Judge Edward J. Devitt held the State of Minnesota to be without authority to regulate the release of radioactive discharges from the Monticello nuclear-power plant. He said:

> The question here is whether Congress has pre-empted the field of regulation of radioactive releases by nuclear-power plants. In my view, it has, and Minnesota is without authority to enforce its regulations in this field.
>
> It should be emphasized that we are concerned here not with the relative merits of the conflicting policies of regulation by the federal and state governments but with the presence or absence of pre-emption of the field by Congress. . . .

Judge Devitt's decision was subsequently upheld by the Eighth Circuit Court of Appeals at St. Louis and by the U.S. Supreme Court.

Several issues were highlighted in the contest between the State of Minnesota and Northern States Power. One was the cost of producing electric power versus the protection of public health; another was the right of a state to impose more protective health regulations in the field of atomic energy than those adopted by a federal agency. Whatever the legal outcome of the Monticello radiation-standards case, politically it was an albatross around the neck of the Atomic Energy Commission and its power-industry partner. It cast the whole Atomic Industrial Establishment into the public-be-damned role of the industrial spoilers of the nineteenth century. In its industry-serving role as promoter of atomic energy, the AEC was now confronted, not merely by a few dissident scientists, but by the government of a state, seeking to protect its citizens and their descendants from the somatic and genetic consequences of a controversial radiation-pollution standard.

The radiation standards adopted by Minnesota had been recommended by a consultant, Professor Ernest C. Tsivoglou, a physicist at the Georgia Institute of Technology. In a report to the state agency, he took the position that the basic issue in the control of radioactive pollution was whether any industrial enterprise has the right to "contaminate the environment beyond the limits of necessity." 3

In asking the AEC for a license to operate the Monticello plant, he said, Northern States Power requested the right to release radioactive wastes to the environment to the full extent allowed by AEC regulations. Such large releases were not necessary, Tsivoglou said, because the industry could build fission reactors which could hold emissions to 1 per cent of the radioactive effluent allowed by the AEC. There was no dispute about this. The industry agreed that the 1-per-cent level of emissions had already been achieved by water-cooled fission-reactor designs. The AEC's director of the Division of Reactor Development and Technology, Milton Shaw, confirmed it.4 The limits adopted by Minnesota, said Tsivoglou, appeared to be both "technologically feasible and economically reasonable."

The state was not asking for zero-radioactive emissions, a goal which seemed to lie in the future, Tsivoglou said, but the necessity for low emission levels was evident in view of plans for two more nuclear-power plants in the region, one at Prairie Island south of the Twin Cities and one to the west.

Under the AEC operating license, Monticello would be allowed an average smokestack release of more than 40,000 curies of radioactivity a day—every day. That, said the Georgia Tech scientist, is "a lot of radioactivity, even if much of it is relatively short-lived." The smokestack releases would contain radioisotopes of nitrogen, oxygen, argon, krypton, and xenon, plus radioiodine if the fuel rods were leaking. Utilities had found it uneconomical to remove leaking fuel rods where leaks were considered "slight" since, as Northern States Power had asserted in its answers to interrogatories, early removal of the rods prevented the utility from reaping the maximum energy benefit from them. In addition to leaks in the fuel-rod cladding, the consultant said, there would be fissioning of "tramp" uranium on the outer fuel-rod surfaces. The Monticello reactor would have about 24,000 fuel rods. Experience showed leaks were bound to develop.

The state permit required the plant operators to minimize fuel leaks and the amount of "tramp" uranium on the fuel-rod surfaces, and to remove iodine from the gaseous effluent. The regulations that Minnesota was trying to enforce set a limit of 0.01 curies a second in the average smokestack release rate, or about 800 curies a day, about 2 per cent of the amount permitted under federal regulations. Tsivoglou said:

> Actual operating experience at other similar plants indicated clearly that liquid and gaseous radioactivity releases can be kept far below quantities that could be permitted under current AEC regulations. The 1968 operating experience in most cases shows these releases to be less than 1 per cent of current permissible releases. In 1968, Dresden I released a total of 240,000 curies of radioactivity in its gaseous effluents, compared to a permissible quantity of 22 million curies. The waste-disposal permit of the Minnesota PCA would allow Monticello a total annual release of 300,000 curies.
>
> There appears to be no reason involving real necessity why such large releases should be allowed, and the Northern States Power Company has indicated in its Final Safety Analysis Report that it expects its actual average release to be about the amount specified in the Minnesota permit.[5]

If that was the case, why did the utility contest the state regulation? It appeared that while Monticello could meet some of the permit limits, the utility did not find it feasible, economical, or necessary to meet all of them all the time. The AEC limits provided a margin of operating flexibility whereby a temporary increase in radiation emission would not force the plant to shut down. But the state standards did not provide this margin. Over and above the impact of the state standards on the operations of the Monticello plant hung the dire possibility that if Minnesota could enforce its own radiation standards, so could every other state. The power industry could visualize fifty different radiation-safety codes, in addition to the federal regulations. The doctrine of states' rights in radiation standards loomed as a serious threat to the whole nuclear-power industry.

On another front, the Minnesota Environmental Control Citizens Association (MECCA), a group of University of Minnesota graduate students, and 8 other organizations intervened at the operating-license stage of the licensing procedure and forced the AEC to conduct a public hearing. The intervenors challenged the safety of the AEC radiation standards and compliance by the company with AEC safety requirements.

A point of basic contention in the hearings was the deletion by the AEC of portions of certain inspection reports which were prepared by the agency's Division of Compliance. The reports normally are public records and the intervenors asked for copies of them in order to determine whether the utility was fulfilling design safety requirements. When they received the reports, the intervenors found that wholesale deletions had been made in the copies which resembled censored dispatches from a military front. Names, times, and places, and in one instance even the title of the report itself, were blotted out in the copies turned over to the intervenors. The deletions for the most part concealed the identities of contractor and subcontractor employees. Prime contractor for the Monticello plant was the General Electric Company. The Chicago Bridge and Iron Company was building the reactor vessel and containment system and the Bechtel Corporation was providing construction, architectural, and engineering services. According to the AEC, the agency's director of regulations claimed the deleted information was "privileged." [6] This view was taken presumably because the agency regarded itself the custodian of contractor proprietary information.

The Atomic Safety and Licensing Board hearing chairman, Valentine B. Deale, a Washington, D.C., attorney, referred the protest of the intervenors at the deletions to an appeals board, which upheld the right of the director of regulations to censor inspection reports when they were submitted to public scrutiny.

The intervenors said they could not fathom what proprietary interest could be involved in these deletions. While the intervenors eventually were able to fill in some of the deleted information in the reports, they were not able to establish the principle that they were entitled to unabridged reports. In a brief they filed with the board, the intervenors charged that there were indications of "several violations of technical specifications."

They charged that the company dragged its feet in making available the results of a leak-rate test in the primary containment structure. The intervenors wanted uncensored inspection and compliance reports in order to ascertain whether safety regulations were being complied with. Names of contractor personnel making the reports were a part of the record. Why could they not see the whole record? The intervenors were not satisfied with the explanation that proprietary interests were involved and were being protected by the AEC. But it was the only explanation they got.

Poring over the specifications of the plant, the intervenors said it was revealed that no automatic shut-off (SCRAM) system was provided in case of an earthquake. Was this not a matter affecting public safety?

"Consistently throughout the hearing," the intervenors said, "the shortcomings of the entire AEC inspection program have been brought forward to this board by way of cross-examination. Slipshod quality-control programs have been indicated on the part of the AEC and little if any actual testing and inspection by the AEC."

Three University of Minnesota graduate students who intervened wanted to know whether the utility was complying with all federal regulations in constructing and testing the plant, in view of what they considered to be inadequate supervision by the AEC. They argued that the public had a right to know this and that the right was being abridged by censored and "sanitized" inspection and compliance reports.

The students were Kenneth Dzugan, George B. Burnett II, and Theodore J. Pepin, all doing graduate work in physics at the university. In a statement filed before the licensing board August 24, 1970, they charged that a "substantial defect in design" existed in the reactor in that it lacked a detection system in the main pressure vessel to detect loose objects that could block the flow of water coolant to the fuel rods. Such blockage could cause the reactor to overheat and melt down. The graduate students alleged also that "substantial areas" of the weld seams in the reactor pressure vessel were not tested.

In addition to the students and MECCA, the eight other intervenors were the Minnesota Committee for Environmental Information; Northern Star Chapter, Sierra Club; St. Paul Trades and Labor Assembly; League of Women Voters, St. Paul; Minnesota Conservation Federation; Minnesota Environmental Defense Council; Clear Air-Clear Water, Unlimited; and the subcommittee on Prevention of Radioactive Pollution of the St. Paul Planning Board.

In several briefs filed with the board, the intervenors noted that actual testing of reactor components was done by the contractor and the subcontractors, rather than by government inspectors or personnel. "At many points . . . it was clear that the Regulatory Staff [of the AEC] does no independent testing, but rather dedicates itself to searching through stacks of paper and taking the word of the applicant." [6]

Moreover, the AEC does not do its own monitoring of the plant site or surrounding area for radiation, but depends on the applicant's data. "This is a terrible responsibility to place upon the Applicant," the intervenors charged. "It first must detect that it is in violation and on the basis of that violation, cause itself substantial monetary loss by retiring the Monticello facility for however short a length of time is involved."

The intervenors expressed doubts about the impartiality of licensing boards because of "their total connection with the AEC." Moreover, since the boards accepted AEC standards it was obvious they could not act on the basis of challenges to the standards. The intervenors criticized the license-hearing procedure:

> . . . on one side of the table is the Atomic Energy Commission with all of its financial backing and on the other side is a utility with assets of approximately $2 billion. Somewhere along the line, the proceedings must be changed so that the public which is neither a promoter nor a regulator nor a utility has the financial resources made available to it so it can make a truly adversary proceeding.

Changes in the proceedings were being considered in Washington, but instead of promoting the public's adversary position vis-à-vis the power industry, the changes were being designed to muzzle critics by reducing public-hearing opportunities and the kinds of environmental issues which the licensing boards could properly consider.

Another view of the AEC and its licensing procedure was expressed by Representative Hosmer of the Joint Committee. "The AEC is not a nuclear Mafia, conspiring to cram unsafe and unneeded nuclear plants down the throat of an unwilling but helpless nation," he told a symposium on nuclear power at the University of Minnesota. He argued:

> The predicted population of 300 million people in this country by the year 2000 will create tremendous demands for electricity. . . . Furthermore, the AEC's licensing process is not a device to boost the sale of nuclear-power plants without regard for the adequate protection of the public health and safety. . . . Nowhere in the world today or at any previous time has there been a more meticulous, detailed, and scrupulously unbiased machinery for assuring public protection against any hazard than that in the AEC's licenses and regulations protecting against radiation hazards." [7]

Hosmer defended federal pre-emption of the field of regulating the releases of radionuclides. "It is inconceivable that any single state could develop either the financial resources or the technical competence to provide the same level of regulatory protection to the public," he said.[8]

The California Congressman cited Tsivoglou's report and the state regulations derived from it as a demonstration in the "shortcomings" of dual, federal-state regulation in an area where only federal authority could command the requisite expertise and detachment.

Tsivoglou had sent a copy of his report to the International Commission on Radiological Protection, and the secretary of the ICRP referred it to H. J. Dunster, deputy division head of the Radiological Protection Division, Health, and Safety Branch, United Kingdom Atomic Energy Authority at Harwell, England, and a member of the ICRP's committee on application of ICRP recommendations.[9]

Dunster approved some portions of the report and criticized others in a letter to Tsivoglou dated August 20, 1969, saying at one point, "Your proposals seem somewhat extreme and could certainly not be related to the recommendations of ICRP." And at another, "I can say categorically that the radioactivity standards you have recommended are not based on ICRP recommendations." Dunster also said he concluded there were some special political problems associated with pollution control or the introduction of nuclear power into Minnesota.

Citing these observations, Hosmer said:

> Dunster's letter reveals how an unbiased world authority looks at the Minnesota Pollution Control Agency's understaffed, underexperienced attempt at regulation. Because of these demonstrated shortcomings of dual regulation in practice, I suggest that responsibility for regulation of nuclear power be left solely with the federal government.

On September 8, 1970, the Northern States Power Company received a limited license to operate but its victory was Pyrrhic. MECCA and other environmentalist groups shifted their campaign to the Minnesota legislature, where bills were introduced to declare a moratorium on the construction of new nuclear-power plants until AEC radiation standards were revised downward or the issues in the radiation controversy were clarified. In a special message to the legislature, April 1, 1971, Governor Wendell R. Anderson recommended "a moratorium on further construction of new nuclear-power plants in Minnesota until such time as the Minnesota Pollution Control Agency certifies that new development can safely begin."

The very next day the Pollution Control Agency received a letter from Robert H. Engels, president of the Northern States Power Company, announcing it would modify the Monticello plant to reduce radioactive discharges.

The modifications consisted of a hydrogen-oxygen recombiner which converts gaseous releases into water and reduces the volume of effluent gas by 80 per cent; an off-gas hold-up system, which would retain radioactive waste gases fifty hours—long enough to allow much of their radioactivity to decay and reduce radiation levels at the plant fence, according to the company, by "not less than 95 per cent"—and finally the installation of an activated-charcoal filtration system in the smokestack, as recommended by the state's consultant, Tsivoglou. Both the recombiner and a charcoal-filtration system had been cited in the utility's District Court complaint and associated statements as two items of extra expense that it would be forced to incur under the Minnesota standards, which it was asking the court to declare null and void. Now it was adding this equipment voluntarily. Engels said that Monticello would be the first large nuclear-power plant in the United States to use recombiner equipment and the first with an activated-charcoal system in the smokestack.

In a letter to the Minnesota Pollution Control Agency, he said:

> Based on the best advice and analysis which could be secured, and assuming installation of the three plant modifications, we are willing to assert that the plant will meet the gross beta-gamma limits for all liquid releases now set out in the Minnesota permit. . . . we can assert with confidence the plant will not exceed the gross beta-gamma release rate limits for gaseous releases now set out in the Minnesota permit.[10]

Engels said he was hopeful of implementing the zero-emission concept of the Westinghouse Electric Company, the nuclear-systems contractor for the utility's next atomic-power plant at Prairie Island.

Grant J. Merritt, the Minnesota agency's executive director, advised Engels that the agency was pleased that the utility "is responding positively to the strong public opinion favoring substantially reduced radioactive emissions at Monticello." He asserted firmly, however, that the state would continue to pursue through the courts "its legal right to set standards in the radioactivity field."

In the context of the radiation controversy, the intervenors and the state had scored an important point. They had succeeded, apparently, in persuading the company to comply with Minnesota's "understaffed, underexperienced attempt at regulation," as Hosmer had characterized it. The state had won a *de facto* victory: A new kind of nuclear-power plant regulation had emerged—

regulation by intervention of the citizens. Or, as utility men put it, by the "mob."

The intervenor effect was even more clearly demonstrated in Michigan where a similar citizens-utility confrontation took place in the spring and summer of 1970. Target of intervention in Michigan was the 700-megawatt (electrical) Palisades Nuclear Power plant, a pressurized-water reactor, built by the Consumers Power Company on the eastern shore of Lake Michigan, about four-and-one-half miles south of South Haven, Michigan.

As in Minnesota the citizens' groups moved to intervene at the operating-license stage of the federal regulatory process rather than at the earlier construction-permit stage. In both cases, start-up of the completed or nearly completed plants was delayed by intervention for months. Under the regulatory process the utility's request for an operating license is granted by the AEC without a public hearing after inspection requirements are satisfied unless someone intervenes and establishes the need for a public hearing. The only notice that such a license is to be granted appears in the Federal Register. Groups in both Michigan and Minnesota complained that restricting the legal notice to the Federal Register, which most citizens never see, has the effect of keeping in the dark the residents of areas where nuclear plants are planned.

Six organizations moved to intervene and demanded a public hearing in the operating-license application. They were the Michigan Steelhead and Salmon Fishermen's Association, TEMP (*T*hermal *E*cology *M*ust Be *P*reserved), Concerned Petitioning Citizens, the American Fishing Tackle Manufacturers, the Sport Fishing Institute, and the Michigan Lakes and Streams Association. These organizations later were joined by the Sierra Club and by a Chicago-based organization, Businessmen for the Public Interest.

In their petition filed with the AEC on April 8, 1970, the groups asserted that the agency's intended action to issue the operating license would endanger public health and safety by producing a radiological hazard, would cause the loss of federal tax revenue, and destroy a recently rehabilitated interstate fishery. It also would cause unnecessary degradation of the environment, significant private-property damage, and loss of tourist revenue.

In addition to building up harmful concentrations of radionuclides in the lake, the petition charged, the Palisades plant's emission of hot water to the lake from its condenser-cooling system would create unwanted algae growth near shore, increase the toxicity of pollutants already present, interfere with the life cycle of some aquatic species, and kill off some fish adapted to cold water.

Noting that the AEC has persistently disclaimed jurisdiction over the thermal effects of nuclear reactors, the intervenors argued that the AEC's responsibility for heat produced by reactors that it licenses was clearly indicated by the National Environmental Policy Act of 1969. The agency had contended that thermal effects were not relevant to cases involving plant licensing under its interpretation of the Code of Federal Regulations, Title 10.

Among owners of property near the plant, Professor John A. Simpson, a cosmic-ray physicist at the University of Chicago, addressed a letter to Chairman Seaborg asserting that thermal pollution from the Palisades plant would accelerate erosion of the lake shore in winter. It would soften or melt the ice which builds up on the shore and protects it against the grinding of loose-floating ice and water churned against the shore by a prevailing northwest wind.

Starting in June 1970, hearings on the Palisades plant's operating permit continued intermittently for eight months before an Atomic Safety and Licensing Board. Like their counterparts in Minnesota and in Pennsylvania, the concerned citizens of Michigan had done their homework on the environmental effects of nuclear power. Mrs. Martha Reynolds, a member of the Conservation Department of the United Auto Workers, said:

> Citizens are becoming increasingly incensed at the absurd contention that they have to choose between power and pollution. People are beginning to speak out against a power industry which on the one hand uses public-relations commercials to promote good will while on the other uses the threats of blackouts and power shortages to frighten the public and conservationists into being quiet, so that the utilities could pollute at will.

Formal assurances were given by both the AEC and the utility that releases of radioactivity would be only a small fraction of the maximum allowed under AEC regulations. In his response to Simpson's letter, which had raised the question of radiological safety as well as thermal pollution, Seaborg had said:

> Very small amounts of radioactive materials generated in the nuclear processes may be released routinely into the environment at controlled rates and in controlled amounts through the cooling water and the ventilation air from a nuclear power plant. . . . Our experience thus far indicates that such releases from licensed nuclear-power plants are kept at a small percentage of even the permissible levels, and that no effect has been discernible on the environment or on the people of these plants.

Of course, Sternglass had claimed a discernible effect in the changes of infant-mortality rates in the vicinity of several nuclear-power plants in Illinois, Michigan, Pennsylvania, New York, and California, but his correlations were routinely rejected by the AEC as unscientific and erratic.

On the utility side, Russell C. Youngdahl, a senior vice-president of Consumers Power Company, issued this assurance:

> Radioactivity will be routinely released from the plant, but in very small amounts which are carefully monitored and controlled. The releases will be within the limits established by the AEC's regulations and will normally be at a very small fraction of those limits.

It was the story that utilities and the AEC were telling all over the country: Don't worry about the AEC standard; we won't come near it—except, perhaps, once in a while. The intervenors didn't buy the assurance, even though it happened to be perfectly true. By the spring of 1970, light-water-reactor technology had reached a level of sophistication where radioactive emissions in gaseous and liquid effluents could be held at 1 per cent of the maximum permissible emission of 500 millirems a year at the plant fence. But the AEC seemed to be clinging to the higher standard. Why? Only pressure by intervenors and questions raised by scientists had elicited the real capabilities of the plants to keep emissions at low levels. There were plants which were "dirtier" than those which could hold emissions down to 1 per cent of the AEC maximum, but which were still under the maximum. The AEC radiological regulations were designed to accommodate the emissions of the earliest of the generation of boiling-water reactors, including some fuel-rod leakage of radiation into the environment and occasional "bursts" of radioactive releases.

This point was driven home by Russell Packard, a Chicago attorney with property near the Palisades plant site. He told the licensing-board members at a hearing:

> Well, you and I have had some experience with arithmetic and we know that annual averages are a flattening of a cycle. This does not prevent at times a high point of the cycle being greatly in excess of the radioactive waste to be emitted into the air and into the water —greatly in excess of that now permissible under the rules. I request as an owner of abutting property that this operating license be denied until a safe maximum is included in the regulations and the applicant is required to adhere to such maximums.

In Michigan, as in Minnesota, intervenors viewed the AEC as an environmental spoiler and regarded its radiation-protection standards with distrust. The standard seemed to imply that there was a "safe" radiation dose, a threshold below which radiation did no apparent harm. In spite of official denial, the effect of the standard in allowing radioactive emissions at levels a hundred times higher than the levels at which the state of the art could keep the emissions suggested a basic affirmation of the threshold concept in AEC regulatory policy.

The threshold concept emerged in testimony September 3, 1970, by Roger W. Sinderman, health physicist for Consumers Power Company, as he was cross-examined by Myron M. Cherry, a Chicago attorney representing all intervenors except the Sierra Club:

> CHERRY: Mr. Sinderman, does your expertise school you in the effects of radiation on human beings?
> A: Yes.
> Q: The Atomic Energy Commission's regulations, are they not, are based on the assumption that there is a safe dose of radiation? Is that correct?

A: The regulations provide intensive—or Part 20—an allowable dose, and in air or water, allowable concentrations not to be exceeded.

Q: Now in your position as a health physicist, does that mean that a person can continually absorb that amount of radiation permitted in the air and water and not suffer any damage or injury because of it?

A: Yes.

Q: Therefore, is it not true, then, that is what we might call an assumed safe dose of radiation?

A: Yes. It is an assumed safe dose.

Q: Have you made any independent studies, Mr. Sinderman, as to effects upon a human being of 170 millirads per year of radiation?

A: Not at that dose level, no.

Q: In your studies as a health physicist, have you come across any investigations with respect to that dose level?

A: Yes, a number of them.

Q: And in your studies were you also—did you also share the opinion that anything below 170 millirads per year was okay for the human body?

A: Yes, I share that opinion.

Q: At what point do you, as a health physicist, as a representative of Consumers [Power Company], become concerned about radiation to a person in a given period of time, say, a year? 171, 172 [millirads]?

A: From a strictly biological viewpoint?

Q: Well, let's phrase it in terms of your agreement that 170 is a safe dose—up to 170. Based on that assumption, what is, in your opinion, an unsafe dose?

A: Recognizing that the normal operation of a plant must produce, or release, radioactivity such that that dose is not exceeded, then one becomes concerned at 171. But from a biological viewpoint, at least ten times higher.

Q: Do you mean that 1700 millirads per year you still wouldn't be very concerned about?

A: That's right.

A professional engineer, W. D. Mohr, of Benton Harbor attacked the siting of a nuclear-power plant on Lake Michigan. He said that in the event of a catastrophic accident that dumped radioactivity into the lake, the water supply of millions of people would be polluted.

There were five basic reasons, Mohr said, why the plant should not be licensed to operate. First, the hearing was dealing with a power plant at a single site without considering other projected nuclear-power plants on Lake Michigan. Second, "there exists in this plant an excessive risk to local population." Third, the risk to the lake from radioactive and thermal pollution was "excessive." Fourth, "the deliberate pollution of Lake Michigan with liquid, radioactive wastes is preposterous." And finally, "the lack of sensitivity of the management of Consumers Power Company for ecological effects of this plant is depressing. . . . Unless and until a national or regional program is developed to oversee the siting of nuclear plants, no license to operate should be granted for anyone."

Pointing out that plans for the plant called for the discharge of liquid radioactive waste along with condenser-cooling water into Lake Michigan via the condenser-discharge canal, the engineer noted that Lake Michigan "is something of the appendix of the Great Lakes system." Its geographical position, he said, makes the lake particularly susceptible as a pollution trap inasmuch as its outflow is restricted to the Straits of Mackinac and the Chicago River-Illinois Waterway system. The average residence time, or flushing capacity, of the lake water is about a century so that slow trends developing in the lake are not diminished greatly by nature.

The intervenors then filed a brief with the licensing board arguing that the AEC's radiation-protection standards not only were inadequate but illegal, inasmuch as they failed to comply with the Atomic Energy Act provision requiring the AEC to regulate atomic energy so as to protect the general welfare and the health and safety of the public. The brief contended that AEC standards failed to do this because:

> They do not take into account radiation doses which the public may receive from sources other than a particular licensee of the AEC.
>
> They do not take into account accumulation of radioactive emissions of radiation by a licensee.
>
> They do not take into account differences in toleration of radiation in different human beings in different places.
>
> They do not adequately provide for a tracing of emission of ra-

dioactivity through all pathways by which the radioactivity could be transmitted to the population.

In view of these alleged deficiencies, the intervenors asserted that the AEC could not legally hear the appellant's request for a provisional operating license "unless and until valid standards were promulgated." They also challenged the legality of the hearing procedure in the light of the AEC's failure to consider thermal pollution of the lake under the National Environmental Policy Act and the Federal Water Pollution Control Act.

So far as thermal pollution of the lake was concerned, it was the understanding of the intervenors that condenser cooling water would be discharged into the lake at a temperature 28 degrees Fahrenheit higher than the surrounding lake-water temperature. The Department of the Interior had recommended a limit of 1-degree rise in temperature of the water being returned to the lake from cooling systems. But the Michigan Water Resources Board had approved a 28-degree rise in temperature in lake water used to cool the condenser of the Palisades plant. This approval was one of the main stimuli to citizen intervention in the operating-license hearing. It triggered the circulation of petitions in which 35,000 Michigan residents protested the plant's operation.

Facing the prospect of long litigation of pollution issues in the federal courts, the company opened negotiations with the intervenors for a settlement of the environmental questions. This move was highly significant. For it conferred upon citizens organizations *de facto* regulatory powers which the federal agencies had failed to invoke. In struggling to impose their conception of environmental law and order in Michigan, the intervenors were acting as environmental vigilantes in the absence of an acceptable code. It was an old tradition in America, glorified by a thousand western movies. When the town marshal failed the citizens acted. In the Palisades case, as in others, the AEC persisted in playing the role of the hapless marshal. The citizens then attempted to enforce what environmental-protection law there was, since the marshal wouldn't do it. The AEC took the position that heat, the principal form of energy released by atomic fission, was not within its jurisdiction. It insisted its writ ran only to radiation, not thermal pollution.

On March 12, 1971, the intervenors and Consumers Power Company announced an agreement of historic proportions in the evolution of atomic power. In return for dropping a court test of the legal issues by the intervenors, the company agreed to modify the condenser-cooling system, already installed, to a closed-cycle system using cooling towers within three and a half years—a process that would greatly reduce the discharge of hot water into the lake. Pending installation of the modified cooling system, the intervenors agreed that Consumers Power could operate the plant at any power levels authorized by its AEC license.

The company also agreed to replace the radiological-liquid waste system already installed with another designed so that "radioactive materials in liquid discharges from the Palisades Plant to Lake Michigan are essentially reduced to zero." However, the agreement provided that radioactive materials in the laundry-waste system could be released to the lake at levels not exceeding 25 picocuries (millionths of a millionth of a curie).

Under normal operating conditions, radioactive materials removed from liquid wastes would be accumulated and shipped to an authorized disposal area. Under unusual conditions, when some radiological waste would be released, the amount would not be more than 10 per cent of AEC limits.

In addition, the company agreed to install a radiological-gaseous waste-control system if a suitable one should become available at an acceptable cost. Meanwhile, the company would support the developmental progress of such a system. Both parties understood that no engineered system was available at the time of the contract negotiation that would reduce gaseous radioactivity from a pressurized-water reactor to essentially zero.

Consumers Power estimated the cost of these modifications at $11 million and it was stipulated that the excess cost must be considered in the utility's base rate for electric power if the agreement was to become effective. This meant, of course, that customers would pay a little more for power from the company as a result of the modifications. No one was giving anything away.

With agreements between intervenors and utilities in Minnesota and Michigan, a new pattern of regulation in the atomic-power industry appeared in 1971. It was a result of a grass-roots revolt against the old, industry-accommodating standards of the AEC. In an upsurge of environmental vigilantism, citizens were using the tactics of delay and the threat of litigation to enforce their demands for environmental and public-health safeguards, bypassing a federal bureaucracy that had been dragging its feet for years.

In negotiations with Detroit Edison during 1971, the Sierra Club and the Businessmen for the Public Interest, the Chicago-based environmental organization, won agreement for the installation of cooling towers and drastic radiological-waste reduction at Fermi II, Edison's big, 1127-megawatt (electrical) boiling-water reactor-power plant at Lagoona Beach, Michigan. Edison agreed to abandon a plan simply to discharge the heated condenser water into Lake Erie and to construct a closed-cycle cooling system instead. The change required the construction of two cooling towers and a cooling pond. The utility also agreed to spend $5 million to reduce the discharge of liquid radiological wastes to zero and to install $10 million worth of special equipment that would cut radioactive gas emission to 1 per cent of AEC limits. These environmental safeguards would cost Edison customers about 1 per cent more for electricity, or less than 15 cents a month on the average household electric bill, the environmental groups' spokesman estimated.

Like Robespierre, who proclaimed he must follow the revolutionary mobs because he was their leader, the Atomic Energy Commission began to follow the intervenor movement. On December 3, 1970, it published amendments to its regulations specifying design and operating requirements that would keep radioactivity in nuclear-power plant effluents "at the lowest practicable levels."

Six months later, on June 9, 1971, the agency spelled out what it meant by the "lowest practicable level." It was the level which was well within the capacity of new atomic reactors to maintain. It was the level sought by the Minnesota Pollution Control Agency. It was below the level which Gofman and Tamplin had for two years repeatedly cited as feasible.

The AEC proposed that nuclear-power plants be designed and operated to limit radioactivity in liquid effluent to amounts not likely to produce radiation exposure in excess of 5 millirems "to the whole body or any organ of an individual" beyond the fence. Gaseous-effluent radioactivity would be limited to a dosage of 10 millirems at the plant fence and beyond, a fiftyfold reduction from the 500-millirem maximum at the plant boundary. In addition, the proposed new guides would limit exposure from concentrations of radioactive iodine and particulate matter to a dosage of 5 millirads off site. The AEC commented in the *Federal Register* announcement of its proposal:

> The specified annual average exposure rates of 10 millirems from noble gases and specified concentrations of radioiodines and particulates at any location on the boundary of the site or in the off-site environment provide reasonable assurance that actual, annual exposures to the whole body or any organ of an individual member of the public will not exceed 5 millirems.

This amount was 1 per cent of the 500-millirem maximum in the existing radiation-protection standards.

The proposed guides, the AEC stated, would provide "reasonable assurance" that the whole body dose to the total population would be less than about 400 man-rems a year. The man-rems figure is the product of the number of persons in a population multiplied by the average exposure, in rems, of each person. If each person in a population of 1 million persons was exposed to 0.001 rems (1 millirem), the total man-rem exposure for the population would be 1000 man-rems.

The AEC qualified its proposal in several ways. First, the new guidelines applied only to light-water reactors and not to any other type of nuclear reactor or facility, such as a nuclear-fuel reprocessing plant, fuel-fabrication plant, or radioisotope-processing plant. The agency explained: "The proposed guides for design objectives for light water-cooled nuclear-power reactors have been selected primarily on the basis that existing technology makes it feasible to design and operate light water-cooled nuclear-power reactors within the guides."

This philosophy of radiological protection seemed to require that radiological safety considerations conform to technological feasibility. The proposal conceded that while "reasonable efforts" were to be made to keep actual releases of radioactivity within the guidelines, "it is necessary . . . that nuclear-power reactors designed for generating electricity have a high degree of reliability." That is, the AEC continued: "Operating flexibility is needed to take into account some variation in the small quantities of radioactivity that leak from fuel elements which may, on a transient basis, result in levels of radioactivity in effluents in excess of the design objective quantities and concentrations."

In this contingency, the Commission *"may* take appropriate action to assure that release rates are reduced" if releases are likely to exceed a range of four to eight times the design-objective quantities, the proposal said. Within this limit, the annual release rates would be expected to keep the off-site exposure rate of individuals within a range of 20 to 40 millirems during a quarter.

Aside from this allowable variation from the proposed standard, which would enable the power plant to keep leaking fuel rods in the reactor for a time, the AEC said the proposed guides would restrict exposure of any individual near the plant to 5 per cent of natural background radiation (averaging 100 to 125 millirems a year in the United States) and exposures of large population groups to about 1 per cent of existing federal radiation-protection guides for the average population dose.

The AEC noted that its proposed guides were consistent with the basic radiation-protection standards and guides recommended by the International Commission on Radiological Protection, the National Council on Radiation Protection and Measurements, and the (now defunct) Federal Radiation Council. The National Council reported on January 26, 1971, that its ten-year study had confirmed the validity of most of the basic radiation-protection criteria used by governmental agencies to regulate the exposure of the public and of radiation workers. These were the familiar standards of 170 millirems per person averaged over the entire population and 500 millirems to individuals in the vicinity of power plants per year. These limits applied to exposures from all sources except medical procedures and natural background radiation. Within this framework, the new limits that the AEC was proposing would significantly reduce the radiation contribution of light-water reactors to the environment by as much as a factor of 100.

It appeared that the citizens who were concerned about radiation and the environment had won an important victory, although Harold L. Price, the AEC director for regulation, asserted that the proposal was not a response to criticism, but had been under consideration for three years.[11] The guidance, the AEC added, took into account comments and suggestions from environmental and conservation groups, states where nuclear-power plants are located, suppliers, utilities, and engineering firms. The agency seemed to have canvassed everybody except the critics within its own ranks. The beauty of the proposed new deal was that any inconvenience to the power industry would be minimal. It would assuage the concerns of the environmentalists without imposing hardship on the reactor designers, builders, or operators, who had three years to get their equipment in shape to meet the proposed guidelines for reactor design and who were allowed to monitor the wastes themselves, on a species of radiological honor system.

At a Washington press conference held on June 7, Price asserted that only three of the twenty nuclear-power reactors then operating were above the proposed guidelines.[12] He identified these as Commonwealth Edison's Dresden I and, possibly, he said, the newer Dresden II reactors at Morris, Illinois, and the Pacific Gas and Electric's Humboldt Bay reactor near Eureka, California. Lester Rogers, director of the Division of Radiological and Environmental Protection, said that these reactors were in the range of four or five times above the proposed limits—but, he pointed out, they were still far below the 500-millirem operating maximum at the plant fence.

The proposed guidelines would affect reactor design and ultimately, if finally adopted by the AEC, would make the 500-millirem limit a redundancy. But they did not change it. The 500 millirems remained the limiting dose to an individual under Title 10 of the Federal Code of Regulations, Part 20, which set "Standards for Protection Against Radiation." The new proposal would amend another part of the regulations (10 CFR 50), pertaining to design objectives and operating conditions for light-water reactors. Thus, under the proposed change, the design standard for a light-water nuclear reactor would be as much as a hundred times more stringent than the radiation-protection standards. Any apparent inconsistency in the two standards was dismissed by the AEC with the statement:

> The proposed guides for design objectives and limiting conditions for operation for light-water-cooled nuclear-power reactors are consistent with the basic radiation-protection standards and guides recommended by the International Commission on Radiological Protection, the National Council on Radiation Protection and Measurements, and the Federal Radiation Council.[13]

In proposing new design objectives, which the industry already had met, the agency appeared to be de-fusing the radiation-hazard argument of the critics of atomic-power plants without yielding to demands for a tenfold reduction in the radiation-protection standard, which remained unchanged.

Several weeks after making that concession to public concern, the AEC moved to blunt the weapon that citizens had been wielding so effectively in Michigan and Minnesota to persuade utilities to reduce radiation and thermal pollution. The agency proposed the amendment of its licensing procedures to prevent public intervention at the operating-permit stage.

The Joint Committee favored the licensing-procedure amendment which would reduce the threat of public intervention as a restraint on nuclear-power plant proliferation. Commented Senator Pastore of Rhode Island, the Joint Committee chairman:

> This is an expanding nation. We've got to have more energy. They're talking about establishing a nuclear plant in my state and I tell you—all hell has broken loose.[14]

6
Boot Hill

A thousand feet below the streets of Lyons, Kansas (population 4503), lie beds of salt 300 to 400 feet thick, deposited with the ebb of epicontinental seas in Permian time, 250 million years ago. Salt has been mined in central Kansas since 1887. It is part of the state's mineral wealth, which includes gas, oil, and the ash of ancient volcanoes that can be processed into slabs resembling finely grained wood for furniture and building materials.

In June 1970, the salt beds under Lyons and its environs were designated a critical resource for the energy future of the United States. The AEC selected them as the site of the first national repository for the nation's rapidly accumulating tonnage of radioactive waste.

Fission reactors produce great quantities of waste radioactivity which must be segregated from the biosphere. These wastes are the products mainly of nuclear fission in the fuel and to a lesser extent of neutron bombardment of otherwise neutral materials in the reactor, including the metal of the reactor itself, coolant fluids, and air and other gases. The mass of radioactive fission products produced in a reactor is nearly as great as that of the fuel that is burned.[1]

Within a reactor, fuel elements where fission takes place are encased in rods of zirconium alloy or of stainless steel. When most of the fuel is used up, the fuel elements consisting of uranium in metallic oxide or carbide pellets are removed from the reactors in their casing and hauled to fuel-reprocessing plants. There the highly radioactive products of the fission reaction (which has produced the heat energy which boils water and makes steam to generate electricity) are separated chemically from the unspent fuel, which is then repackaged into new fuel elements. These are shipped back to the power reactor, leaving a variety of radioactive isotopes to be disposed of. This high-level waste is concentrated in a liquid or slurry form and is stored in steel or steel-and-concrete tanks for up to five years to allow much of its radioactivity to dissipate. The liquid wastes are not only radioactively "hot," but also thermally "hot," generating heat at densities up to 200 watts per gallon. It is this combination of radioactive and thermal energy in the high-level wastes that has created a storage and disposal problem unique in the history of mankind. Because they are so highly radioactive, the wastes cannot be allowed to contaminate the biosphere; because they are thermally hot, it is difficult to keep them contained. They emit large quantities of beta particles and gamma rays, and they are long-lived so that they must be kept out of the environment of man for 500,000 to 1 million years.

During the first quarter century of the atomic age, high-level liquid wastes were stored in tanks at AEC reprocessing plants at Richland, Washington; at Arco, Idaho; and on the Savannah River near Aiken, South Carolina. Leakage from the tanks and into the ground was persistent. In a report in 1968, the General Accounting Office cited leakage of 227,000 gallons of the 93 million gallons of wastes stored at Richland and at Aiken. Most of the "hot stuff," about 74 million gallons, was tanked at Richland, where leaks were found in 10 of 149 underground storage tanks. There was no provision for secondary containment and the radioactive poison, emitting up to 10,000 curies per gram, simply percolated into the ground where it threatens to contaminate the water supply for millennia to come.

In South Carolina, where 17 million gallons were stored at the Savannah River plant of the AEC, leakage was found in four tanks, but in only one instance did high-level waste seep into the ground. In Idaho, all 1.6 million gallons were contained. An investigation by the Federal Water Quality Administration (FWQA) in the spring of 1970 complained that the practice of burying solid radioactive wastes in trenches and dumping low-level wastes into pits and ponds at the National Reactor Testing Station site had degraded the ground water under the station, located near the eastern edge of the Snake River Plain in Southeastern Idaho. This region is underlain by the Snake River Aquifer, one of the world's most productive underground-water reservoirs. The FWQA recommended that the station stop burying radioactive wastes, disinter those it has buried, and no longer dispose of chemical wastes in seepage ponds in order to protect the underground water supply.

The AEC realized that these storage devices were unsatisfactory and could never be tolerated as long-term solution to the problem of mounting radioactive waste. Since 1959, the agency had been developing methods of solidifying the liquid wastes by evaporation. But while the process simplified waste control, it did not fulfill the long-term requirement of getting rid of the high-level wastes permanently.

In addition to the high-level waste, another type called "alpha" wastes had to be disposed of. These are materials, including articles of clothing and equipment, contaminated by small quantities of plutonium and other transuranium isotopes in the course of nuclear-fuel processing. For the most part, these wastes consist of short-range emitters of alpha particles (helium nuclei) which do not generate much heat. The alpha wastes are less difficult to store but present a similar requirement of long-term segregation from man's habitat.

With the anticipated increase in nuclear-power production from 6000 megawatts in 1970 to 940,000 megawatts by the year 2000, the amount of wastes from fuel-reprocessing plants could be expected to increase eighty times, the AEC estimated.[2] In addition to the three AEC reprocessing facilities at Richland, Arco, and Aiken, two private reprocessing plants have been constructed, one in New York State (West Valley near Buffalo) and the other at Morris, Illinois. More will probably come with the advent of breeder reactors in the 1980s.

The question of what to do with the mountain of radioactive

waste that the expanding fission-reactor industry is producing was not new to the AEC. The problem had been studied and debated, with the assistance of two committees of the National Academy of Sciences-National Research Council, since 1955. It was the consensus of all the studies that salt beds deep underground offered the best prospects for long-term, safe, and undisturbed storage. The need for a repository was critical, for the poisonous refuse that the atomic power industry was creating threatened the future of life on this planet. Lofting the wastes into solar orbits or dumping them on the moon would be too expensive.

Salt beds lying below the underground water supplies of the United States had a number of advantages as natural containers for the waste. Salt withstands very high temperatures, and when heated becomes plastic and flows, thus sealing fractures and filling holes. Its compressive-strength and radiation-shielding properties are comparable to those of concrete.

Three potential sites where salt beds underlie the surface at depths of 500 to 1500 feet were considered by the AEC.[3] One was in west-central New York. A smaller area lay underground in southeastern Michigan, mostly beneath the Detroit metropolitan area. This prospect seemed to be a remote one for political, if for no other, reasons. A proposal to bury wastes with an accumulated radioactivity on the order of 270 billion curies beneath an urban center of 3 million people was not a gambit calculated to win popular support for the AEC.

"On balance," the AEC reported, the area around Lyons, Kansas, appeared to be the most favorable.[4] It is a thinly populated, rural district, but is accessible by both road and rail. Moreover, the Oak Ridge National Laboratory had been experimenting for several years in a salt mine at nearby Hutchinson, Kansas, to determine the effects of thermally and radioactively "hot" wastes on salt and had found the results quite promising. The salt bed made a seemingly ideal repository.

In the abandoned mine of the Carey Salt Company at Hutchinson, radiation emitters simulating radioactive wastes and heaters had been emplaced in salt to create effects of actual wastes in a repository. The mine was bedecked with instruments for recording temperature and radiation, and their mechanical effects on the salt.

The AEC thus proposed purchasing 1000 acres of land just north of Lyons to develop the waste-disposal facility over a period of years, first demonstrating that the burial scheme would work in an abandoned salt mine under the town of Lyons. High-level wastes ready for burial would be hauled to Lyons in railroad cars, after being solidified at the reprocessing plant into a ceramic matrix and encased in stainless-steel cylinders 10 feet long and 6 inches in diameter. Delivered in huge casks to the radioactive "dump" at Lyons, the cylinders would be taken into a salt mine or "room" excavated in the underground salt beds and lowered in holes drilled into the salt.

About twenty cylinders could be emplaced in a room 30 by 300 feet. The holes would be sealed with salt. Then the entire room would be backfilled with the salt excavated from it. Researchers at the Oak Ridge National Laboratory had estimated that the twenty cylinders would produce an ambient temperature of 200 degrees centrigrade. When the room had been backfilled and sealed, the accumulating heat would cause the fill to flow plastically and merge with both floor and ceiling, thus closing the "salt vault" forever—or at least for hundreds of thousands of years.

The burial in the salt would be considered permanent, mainly because, after several years, it would be impossible to retrieve the wastes and move them somewhere else. Under the corrosive action of the salt and the effects of the radioactivity and heat of the waste, the stainless-steel cylinders would disintegrate in about six months and the ceramic matrix in about three years. After ninety years, about 95 per cent of the backfill would have recrystallized, sealing the radioactive tomb.

Oak Ridge played a leading role in the investigation and testing of salt for radioactive-waste storage. The laboratory's director, Alvin M. Weinberg, had no doubt that salt was the way to go in disposing of this treacherous refuse. He characterized the AEC's selection of salt beds for a repository as "one of the most far-reaching decisions we or, for that matter, any technologists have ever made." [5]

"These wastes will be hazardous for up to a million years," Weinberg said. "We must therefore be as certain as one can possibly be of anything that the wastes, once sequestered in the salt, can under no circumstances come into contact with the biosphere."

In June 1971 the Joint Committee on Atomic Energy recommended to the Congress that $3.5 million be appropriated for the purchase of land at Lyons. The agency proposed to launch a demonstration that would prove out the determination that Oak Ridge had made that rising wastes of the Atomic Industrial Establishment could be safely and economically interred in the Kansas salt beds.

So it was that Lyons, Kansas, was fated to become the Boot Hill of the atomic age. Lyons is the seat of Rice County which is as flat—and in July likely to be as hot—as a frying pan. The red-brick courthouse, built in 1910, is partly air-conditioned nowadays, but the sweep fans that were hung near the ceilings in summer to cool it are still stored in the attic. The main floor of the silent, old building is a museum of exhibits: Kansas barbed wire, circa 1870; stone arrowheads made by a people called the Quivirans of Caddoan linguistic stock; Civil War swords and English cap-and-ball rifles captured from Confederate raiders. Outside, on a pedestal, a statue of a Union soldier cut in stone stands with his rifle at parade rest, above the motto, "Lest We Forget."

On a hot July afternoon in 1971, the population density of Lyons seemed to be about what it was in 1910. Only an occasional pedestrian was visible out-of-doors in the business section around the courthouse square. A panel truck moving slowly along Commercial Street had the thoroughfare all to itself. In this part of Kansas, the population explosion seems fictitious. The population is declining. In 1960, the census listed 14,025 souls in Rice County. By 1971, the number had dropped to 12,244. Yet the region is rich agriculturally, with an assessed valuation of $69 million for real and personal property. Real estate, mostly farm land, is assessed at $31 million and, since the assessment runs about one-third of market value, the real property in the county was worth about $100 million at 1971 prices.

"Right here in the Arkansas River Valley, good, rich land sells for $300 an acre," said Arthur Harvey, the county clerk.

How did the prospect of living atop a dump that might contain 770,000 cubic feet of solidified, high-level, radioactive waste only thirty years hence affect the people of Lyons?

"None of us is afraid of that," said the clerk. "Some were misinformed and thought they were burying explosives, but as soon as that was straightened out, no one was worried about it at all." In fact, some people thought the repository might stimulate business and attract new people, or least provide new jobs so that young people didn't have to leave the community to make their way in the world. The steady decline in population, Harvey added, could be blamed on the increased mechanization of farming, which seemed to require more machinery and fewer people every year.

Not everyone shared the county clerk's view. Some of the farmers who stood to lose their land when the AEC purchased its first 1000-acre site for initial development of the repository were apprehensive, unsettled, or angry.

Most townspeople in Lyons supported the project, however, in the hope it would stimulate business. W. W. Chandler, Jr., banker and head of the Nuclear Advisory Council of the Chamber of Commerce, predicted that the repository would give Lyons a much needed boost in economic growth. Even though the program would employ no more than 200 persons, the project would attract some kind of industry. The Hutchinson *Daily News* polled 100 persons in Lyons on the desirability of the repository. The poll reported 67 for it, 22 against, and 11 undecided. Chandler and Walter Pile, manager of the Chamber of Commerce, expressed what they believed was the predominant opinion in Lyons on the safety of the project: the AEC could be relied on to safeguard the health and welfare of the community. There was, therefore, no reason not to welcome the repository to Lyons and persuasive reasons to support it.

It did, indeed, appear to the Oak Ridge scientists that the salt beds underlying central Kansas were ideal for waste storage. The beds must certainly be among the most stable and unchanging features of the earth's crust. The salt could never have been in contact with ground water or it would have been washed out. The great beds lay in a geologically calm region where earthquakes were rare and only mild shocks had ever been recorded. At a depth of 1000 feet, the salt beds were hardly likely to be exposed by water erosion. They were farther south than the farthest extent of the Pleistocene ice sheets. The possibility that the radioactive wastes would be uncovered during their long, hazardous lifetime by natural processes operating from the surface seemed most remote.

But the AEC's optimism that burial in salt was the answer to the disposal problem was not shared by the Kansas Geological Survey in its entirety. The survey's director, William W. Hambleton, was not satisfied that the work which Oak Ridge had done in its experimental program called Project Salt Vault in the mine at Hutchinson and in other studies, guaranteed safe containment of the wastes in salt. It was true that the salt beds were sealed off from the Kansas water table above them by tons of rocks, barring contaminants from percolating into the underground water supply. But to the men of the Kansas survey who were mainly geologists and geophysicists there was nothing immutable about this arrangement. In this respect, their outlook differed from that of the Oak Ridge physicists and chemists. The Kansas scientists could not discount the possibility that thermal expansion or gaps in backfill after it recrystallized and solidified might allow the overlying rocks to crack or subside. Fissures would then open up and ground water—even surface water—might seep down into the salt through such fractures. The salt beds had been stable for eons, and they had been let alone for eons; but how stable would they remain when subjected to intense and long-term heat and ionizing radiation? If water entered the salt, a species of radioactive hell would break loose under Lyons and a good part of central Kansas. The water would be heated by the thermal energy of the wastes, which would become free of its containers, free in the salt, in only three years. Thermal convection would start the water flowing through the salt and a highly radioactive salt solution would begin to permeate the aquifers of Kansas and poison the underground water supply for thousands of centuries.

Hambleton, a geophysicist, shared the belief of many of his colleagues that the Pleistocene ice ages are not over—the ice will probably come again. While it had never reached as far south as Lyons, the Nebraskan and Kansan glaciations had surged southward from the Arctic Circle as far as northeastern Kansas. Who could say how far the ice sheet would come down from the north next time? How fantastic this sounded in terms of the A.D. 1971 hopes and ambitions of people in a small Kansas town with a declining population. And, yet, the last of the Pleistocene glaciations, the Wisconsin, has been gone hardly 10,000 years. The advances and retreats of the ice left us Niagara Falls and the Great Lakes, as we know them today. The European counterpart of the last ice surge retreated so recently that the land in Scandinavia is still rising from the relief of the ice load that lay upon it within the oral memory, if not the written record, of man. Who could say that an ice sheet that could sculpt the Great Lakes might not gouge the salt beds of central Kansas out of the earth and dredge up a massive concentration of radioactive waste to the surface? Who could say there would not be another advance of the polar ice in the next 25,000 or 250,000 or 500,000 years—when the wastes would be still "hot"?

More immediately, Hambleton was critical of heat-flow studies made by Oak Ridge scientists. The work did not satisfy state geologists that thermal expansion in the burial vault would not weaken or crack the overlying rock seal. The survey regarded an explicit answer to this problem as "crucial to the safety of the repository," Hambleton stated in a report to Kansas Governor Robert Docking.

Samples of salt exposed to gamma radiation showed the ability to store 10 to 50 calories of heat per gram. Samples irradiated with protons fired from a Van de Graaff accelerator stored up to 80 calories. What would happen if vast multiples of this energy in thousands of kilograms of wastes suddenly were released?

The survey report speculated: temperature would rise as high as 620 degrees centigrade. Salt would flow readily and rapid expansion might cause an explosion that could produce faults in the overlying rocks or force the containers deeper into the salt beds and down into the underlying layers of shale. In view of the fact that an enormous amount of radioactivity would be concentrated in the salt, the Kansas men said, the AEC and Oak Ridge "have exhibited remarkably little interest in studies of radiation damage."

"The State Geological Survey regards this problem as extremely critical to safe storage of radioactives at the Lyons site," the report added, and the problem had not been resolved by Oak Ridge. Nevertheless, the AEC insisted on going ahead with its demonstration program on a "let's see what happens" basis.

Some of the farmers were concerned about the safety aspects of the project and about their being dispossessed. One was Roy Dressler, who owned 160 acres just north of Lyons and raised wheat and milo on it and on a leased 80-acre tract down the road a ways. He and his wife had farmed in the neighborhood for thirty-seven years. In the fall of 1970, they were visited by a real-estate appraiser from Topeka who looked at the property, indicated that the government might want to buy it, and departed without saying much more. After that, they were occasionally telephoned by news services and radio-station reporters, asking them what they knew about the AEC's plans.

"We said we knew absolutely nothing," Dressler said. "Nobody has told us anything. They all seem to think our land is going to be taken. But we just don't know."

Mrs. Dressler, a former nurse, said she was concerned about the radioactivity hazard that the project would import into a region where citizens had adapted over the years to common, everyday ones—cyclones, drought, inflation, recession. Nobody needed radioactivity to worry about.

"If the people of Rice County were allowed to vote on it," Mrs.

Dressler said, "I'm sure this project would not carry."

In his late fifties, Dressler said that he and his wife face a serious crisis if they are forced to relocate.

"We couldn't afford to," he said. "We're not old enough to retire nor do we have enough money and we're really too old to start a farm operation again, because it takes quite an investment to do that. We will not recover our investment here, I feel."

"Physically, it's something for you to start a complete new business," said Mrs. Dressler. "Men don't do that when they're in their late fifties. That isn't when you start a new operation. And even if they gave you money and told you to build some place else, by the time you did and put it in operation, you'd probably be dead and buried."

At the end of June, the Chamber of Commerce sponsored a public meeting on the salt repository at the high-school auditorium. Only sixty persons showed up. The farmers were out that evening cutting wheat, working in the fields until ten or eleven o'clock, Dressler said. His own farm hands never got through until then.

Though the crowd was small, the *Lyons Daily News* reported that the meeting was lively. The controversy between the Oak Ridge and the Kansas Survey men surfaced, but much of its relevance to their safety was lost on the audience, which tended to regard the dispute as academic. The Kansans criticized the Oak Ridge work as "simplistic" and "crude," according to the *News* report. John Halepaska, the state hydrologist, suggested that while the Oak Ridge investigators might have received good answers from the tests of their laboratory models of the salt formation, he and other survey officials doubted that the models were accurate representations of the formation. Director Hambleton insisted, "The numbers that the AEC is bandying around are not real things. They are based on a model that is wrong. That is what John [Halepaska] is saying."

The differences between the AEC and the Kansas Geological Survey dated back to March 24, 1970, when representatives of the AEC and the Oak Ridge National Laboratory met with Kansas Governor Robert Docking to discuss plans for the repository. A month later, the survey advised the Governor of its concern about the inadequacy of existing geological data on the site, and urged that any definite selection of the Lyons site should wait for the outcome of detailed studies. But the AEC moved swiftly to implement its plans.

The AEC had based its interest in the Lyons area on its Salt Vault Study in the Hutchinson mine, which had been carried on with simulated wastes from 1965 to 1967. In the light of the survey's concern, Director Hambleton was invited to serve as a member of the Panel on Disposal in Salt of the National Academy of Sciences Committee on Radioactive Waste Management, which had been appointed at the request of the AEC to advise it on waste-disposal problems. At a meeting at Oak Ridge, Tennessee, in May 1970 the panel was briefed on the design specifications of the repository, which had been drawn by Kaiser Engineers, Inc., and the Oak Ridge scientists. The Kansas Geological Survey was asked to recommend specific studies that would show the feasibility and safety of sinking large amounts of high-level radioactive waste in the underground salt beds. The survey readily complied and submitted its recommendations on June 8, 1970. In the following week, at the request of the NAS Panel, the survey arranged a meeting at the University of Kansas, at Lawrence, to review the problems of salt storage. While these discussions were continuing, the AEC called a press conference on June 17 and announced the selection of the site at Lyons for a demonstration of waste storage in salt, leading to permanent development of a national radioactive waste repository.

This was not the first time the AEC had bypassed or ignored an advisory committee on radioactive-waste disposal. In 1955, when this question became sufficiently worrisome, the AEC asked the National Academy of Sciences-National Research Council for advice. A Committee on Geologic Aspects of Radioactive Waste Disposal, composed of leading earth scientists in industry and academia, was appointed. It began its career in April 1955, as a steering committee, with the principal function of assisting the AEC's Division of Reactor Development and its agents in their search for safe areas for burying nuclear wastes. In addition to burial in salt, the committee considered the feasibility of injecting radioactive wastes as a kind of grout into man-made fractures in shale.

From time to time, the committee criticized the actual practice of waste disposal at the Oak Ridge National Laboratory, the National Reactor Testing Station in Idaho, and the Hanford Atomic Products operation in the State of Washington, where it found that intermediate and low-level liquid wastes were being dumped into seepage pits. The committee pointed out that this method risked contamination of the environment.

At the Idaho Testing Station, pipes had been laid underground without "ordinary safeguards" against corrosion on the assumption the pipes would not corrode in the dry soil—which they did, the committee said. Plutonium wastes with a half-life of 24,000 years were given shallow burial in ordinary steel (not stainless) drums on the same incorrect assumption. The committee warned that ultimate leaking of the plutonium from corrosion of the drums was inevitable and that considerations of long-range safety "are in some instances subordinated to regard for economy of operation." Also, some disposal practices reflected an "overconfidence in the capacity of the local environment to contain vast quantities of radionuclides for indefinite periods without danger to the biosphere."

The committee summarized these findings in a report which it submitted to the AEC in May 1966 after ten years of surveillance. "None of the major sites at which radioactive wastes are being stored or disposed of is geologically suited for safe disposal of any manner of radioactive wastes other than very dilute, very low-level liquids," the committee concluded. A probable exception to this was the grout injection of liquid wastes into shale fissures in the vicinity of Oak Ridge. It added; "the safety and needs of populations in centuries to come demand that methods or facilities be developed now to receive and contain safely in perpetuity the volumes of wastes that will be produced as the nuclear industry expands. . . ."

Following this report, the committee was disbanded. The report itself was withheld for three years by the AEC and was not released until Senator Frank Church of Idaho asked Chairman Seaborg to give him a copy on October 7, 1969.

Why was the report kept in the house? Seaborg explained that the committee "went beyond the requested appraisal of AEC research and development program on radioactive-waste disposal." Also, it commented at length on operations and activities not under the jurisdiction of the Reactor Development Division "concerning which the committee has been given only a short briefing incidental to the research and development program appraisal." The AEC staff, moreover, "had many criticisms concerning the committee report" and these were not resolved before the committee was dissolved in 1967—"preparatory to the establishment of another advisory committee with a broader spectrum of scientific disciplines."

Subsequently, the Committee on Radioactive Waste Management was formed in 1968 "to review and advise the AEC concerning long-range radioactive-waste management plans and programs for an expanding nuclear energy industry."

This was the committee whose deliberations were so expeditiously bypassed in June 1970 and whose eventual recommendations were virtually ignored in the controversy between the AEC and the State of Kansas.

The AEC had made its decision to move ahead with the repository at Lyons, or at least with a demonstration of it, before the Panel on Disposal in Salt could even draft its report to the National Academy committee. On July 22, the AEC invited accredited news correspondents to come to Lyons on July 29 for a tour of the Carey Salt Mine which the agency planned to use for its demonstration project. A full briefing was arranged for the newsmen at the Lyons Country Club, where they received AEC information kits describing the project and badges to admit them to the salt mine. All this was going on while scientists at Lawrence were still planning studies that might tell whether the project was safe or not.

Spokesmen for the AEC hailed the decision to "go to salt" as one of the most crucial of the atomic age. The underground salt beds of Kansas offered a seemingly happy solution to the nagging problem of atomic-waste disposal. But the Kansas survey wasn't happy about it. It became perfectly clear to Director Hambleton and to many members of the Kansas legislature and of the state's congressional delegation that the AEC was moving into Kansas like Quantrill's raiders. Whether Kansans liked it or not, they would have the privilege of living atop the first nuclear garbage dump in history.

In a preliminary report to Governor Docking on project studies, Hambleton warned that despite the AEC's decision to charge ahead, problems of heat flow and surface subsidence remained unsolved. Oak Ridge was to have looked at them in Salt Vault and other studies. It had done so, but the models the scientists used, Hambleton said, were based upon a rock section consisting of a unit of pure salt and a unit of pure shale. Instead, Hambleton said, the actual rock section consisted of laminated salt and shale. There was a difference in heat and radiation effects between the model and the actual section.

No one really knew how much heat the radioactive wastes would emit after they were buried in the salt. The heat-flow calculations made at Oak Ridge were based on assumptions, Hambleton said, and the actual values, which had not been determined, could be—and they *should* be before this project got off, or rather under, the ground.

Hambleton and his survey scientists regarded the problem of heat flow as central to the safety of burying the wastes in salt. He commented rather tartly in the report:

> Oak Ridge National Laboratory and AEC staff have exhibited remarkably little interest in the heat-flow problem and have not demonstrated capability for solving three-dimensional problems involving a complex laminated section. The interaction of subsidence, thermal expansion, and heat flow could be responsible for breaking the seal of overlying rocks and permitting entry of surface or subsurface waters. The State Geological Survey regards solution of this problem as crucial to the safety of the repository site.

In addition, Hambleton reported, high temperatures caused by irradiation of the salt could produce explosions cracking the overlying rocks. Moreover, with the disintegration of the waste containers, radioactive particles could migrate through the salt toward the rocks above or below the beds. Some water is available in the salt, and the particles could be suspended by turbulent boiling.

Furthermore, the report stated, the solid particles free in the salt would expose it to "significantly higher radiation doses." Although the total expected dose per container is 2×10^{10} (20 billion) rads, "we have indications that the dose may be significantly larger." Gamma radiation, the report continued, can cause chemical breakdown of salt. Radiolysis could result in the formation of new chlorine compounds that are capable of leaching plutonium.

On March 16, 1971, Hambleton appeared as a witness before the Joint Committee. He repeated his complaints that the AEC exhibited little interest in the heat flow and radiation effects on the integrity of the salt beds.

There was no question that the Oak Ridge scientists were fully capable of making adequate tests. The fact that they did not, the Kansans surmised, could be attributed to an AEC economy "kick," which limited funds for testing. Until adequate tests were made, Hambleton said, the survey would oppose the repository, with the support of the state administration of Governor Docking.

Essentially the state scientists were in fundamental disagreement with the AEC concept of demonstrating a partially tested system to detect the "bugs" in it—the way all mechanical inventions had been developed, from the steam engine to the Lunar Rover. You tested a device up to the point where you hoped it would work, and then you tried it out to see where it would fail. Survey men did not buy that testing philosophy for the repository, as a failure could cause calamitous consequences to the environment. If the overlying rocks cracked and water poured into the radioactive salt beds; if the surface slumped and fissured; if the water boiled underground and became steam under pressure—radioactive geysers might erupt at Lyons. One could imagine a Kansas version of Old Faithful, blowing alpha particles and gamma rays into the biosphere, and leaving the ground covered with a film of radioactive salt. Could it happen? No one knew. That was the point. No one knew—not on the basis of the AEC's testing program.

The "let's see how it works" demonstration philosophy had characterized the whole development of fission reactors, starting with Shippingport and Dresden I. Fermi I, the first breeder reactor, was a demonstrator. Its coolant was accidentally blocked and its core nearly melted. Five years passed before it could be started up again. There were demonstration reactors supplying power years before effective measures to prevent a core meltdown could be devised. Meltdown was catastrophe. It would happen if the flow of coolant was blocked and it posed the threat of scattering radioactive debris all over the landscape. Under the AEC's reactor-development policy, the machines were demonstrated and put into commercial use long before an emergency core-cooling method was invented or even thought necessary. It was like demonstrating a new model automobile, with brakes to come later.

The AEC took the position in debate before the Joint Committee during the spring of 1971 that the Oak Ridge tests justified going ahead with a demonstration of waste interment in which a limited amount of the high-level radioactive material would be buried in the abandoned Lyons salt mine and monitored to see what happened. If unexpected hazards appeared, the wastes could be retrieved and the demonstration halted, the AEC said. No one explained how retrieval would be accomplished if the wastes containers disintegrated before retrieval was attempted. But seeing what would happen was not a gamble the Kansas survey wanted to take. Since it was possible to do enough experimentation to eliminate the gamble, the survey insisted that the AEC spend the money to do it.

The Kansans perceived a penny-wise, pound-foolish policy in the AEC's reluctance to meet the state's research requirements. A conspicuous example of this policy was demonstrated in the fall of 1970. Under AEC auspices, the United States Army Corps of Engineers moved in a drilling rig from Mobile, Alabama, and cut a 1300-foot core out of the northwest corner of the Lyons site for analysis. The approximate cost of this operation was $125,000. Six inches in diameter, the core was a cylindrical section of a billion years of the earth's past, pulled out of the ground by the hollow drilling tube. It penetrated the recent topsoil and silty terrace deposits from the earlier Pleistocene, and sank into Cretaceous sandstone, red shale, dolomite, and blue shale, perforating the Hutchinson salt beds at a depth of 795 to 1100 feet. Below the salt, the drill cored out more shales. Beyond its reach lay ancient limestone, and finally, at 4000-feet depth, the pre-Cambrian basement of mid-continental North America. The core provided a magnificent, three-dimensional experiment in which the effects of heat transfer at the salt-shale interface, as well as of pressure and radioactivity, could be tested.

The core was shipped encased in plastic-wrapped segments to the University of Kansas at Lawrence for testing. But the AEC ignored it. The core was allowed to remain in an unheated building over the winter of 1970–1971, exposed to alternate freezing and thawing which destroyed much of its value as experimental material, according to Ernest Angino, assistant survey director. The AEC would not provide funds to buy a heater for the building and the Kansas survey did not have money for this purpose.

The episode of the core was widely discussed at Lyons. It seemed to exemplify federal-agency bumbling. It also suggested an indifferent approach by the AEC to the question of safety.

The Dresslers knew all about the core. "Well," said Roy Dressler, "if you could have been here and taken some pictures you could have believed it. The Army Engineers did it. They had so much equipment in here they had to rent ground to park it on. I farmed all around it and I know. I saw them dig post holes with a piece of equipment that must have cost $75,000. It was used to dig ten or twelve post holes just to put a fence around it. We've got lumber yards all over Kansas and two or three in Lyons. They shipped lumber up all the way from Alabama. Real economy!"

The position of the Kansas survey and its concerns were supported by the Department of the Interior. Assistant Secretary Hollis M. Dole advised the AEC that:

> We believe that additional significant studies and confirmatory data concerning the geology and hydrology of the salt deposits and overlying rocks at and near Lyons, Kansas, and the effects of construction of the waste-disposal facility will be necessary to demonstrate conclusively, beyond a reasonable doubt, that these deposits are indeed suitable for the final repository.[6]

The Radiation Office of the Environmental Protection Agency complained that it was not possible for its staff to assess fully the environmental consequences of the proposed repository on the basis of the AEC's Environmental Statement:

> . . . we do not believe the Commission has adequately developed or referenced in the environmental impact statement the results of the work it has done on this problem and which presumably relate to the conclusions contained in the impact statement.[7]

In its budget for fiscal 1972, the AEC asked for an appropriation of $3.5 million to purchase land for the repository and commence the demonstration at Lyons. It requested an additional $21.5-million authorization to develop the repository as a permanent facility. At budget hearings in March 1971, Representative Joe Skubitz of Pittsburg, Kansas, appeared before the Joint Committee to oppose the repository. Although his congressional district, the fifth, did not include Lyons, Skubitz was concerned as a Kansan about the safety of the project and the apparent intention of the federal agency to impose it on Kansans whether they wanted it or not.

"If this Committee authorizes the funds and permits the AEC to purchase the ground, it will have effectively denied Kansas people any choice in this vital issue," Skubitz said. He added that forty-nine members of the Kansas legislature had sponsored bills at Topeka calling on Governor Docking, President Nixon, and Congress to stop the repository.

Hambleton, representing the Governor, told the Joint Committee of Docking's request that funding of the project be deferred until "adequate study and evaluation of the questions and concerns have been completed," adding:

> He now reluctantly concludes that the efforts of the AEC to minimize the problems raised by scientists in Kansas and to treat these concerns as negligible and trivial, support fears of many Kansans that if funds are appropriated for design and site acquisition the project cannot be stopped at a later date if it is indeed found to be unsafe.

Early in April 1971, AEC representatives agreed to define a systematic program of investigations of points raised by the Kansas Geological Survey. Floyd L. Culler, deputy director of the Oak Ridge National Laboratory, said that "a detailed investigation of surface and subsurface geology and hydrology of the site and its environs is under way." [8] In addition, he said, special investigations were in progress to evaluate the probable long-term effects of climatic changes on surface erosion and denudation of the salt bed; to determine the rate at which the eastern boundary of the salt is being dissolved by ground waters; and to locate and seal all existing wells in the vicinity which penetrate the formation. Beyond this, an "extensive study" of the geothermal effects of repository's operation was in progress—"expanding greatly our earlier favorable calculations." [9]

Some members of the Joint Committee were not impressed by the Kansas pleas to go slow in developing the repository. After listening to the arguments, Representative Hosmer remarked: "I get the impression that we should never have invented the wheel if we had thought about it beforehand."

Evidently, the Joint Committee did not elect to think about the problem too long. It recommended the $3.5-million appropriation in the general AEC budget request of $2,321,187,000 for the 1972 fiscal year to buy and set up the demonstration at the 200-acre Carey Salt Mine in Lyons and an additional 800 acres northeast of it.

The AEC appropriation bill came up in the House of Representatives on July 15, 1971. It was presented by Representative Melvin Price of Illinois, who advised his colleagues that questions and objections raised about the repository "cannot be answered until small amounts of high-level radioactive waste to be supplied from government sources are actually deposited in the salt mine to demonstrate the feasibility of using this facility for its intended purpose." [10]

Price added that the Joint Committee agreed that the research should be done and that all reasonable questions raised by Kansas and federal officials would be answered before the depository was put into full operation.

Representative Skubitz then moved to amend the appropriation bill by striking out the repository appropriation. Citing the AEC's rationale for going ahead with a "let's see how it works" demonstration, he said: "Mr. Chairman, this is the contention which is vigorously opposed by the Governor's scientific advisers."

Representative Hosmer defended the AEC-Joint Committee effort to launch the project immediately. He explained that in the original request, the AEC had asked for $25 million in construction funds to build the repository. Hosmer said:

> In deference to the gentleman's fears and alarms and so on, that request was scaled down in the authorization to $3.5 million. . . . What this authority seeks to do here is to permit the AEC to acquire this property so that research and development can continue to determine whether ultimately, in fact, it will be safe for all time to store this high-level radioactive material in this particular place. If so, the title to the property will be secure and the storage can proceed then, not now . . . but then. Let us not place the Government of the United States in a corner where it can be held up and robbed.
>
> SKUBITZ: Does the gentleman admit that this is an abandoned salt mine and has been abandoned for about fifty years?
>
> HOSMER: I do admit that it has been in existence for over fifty years.[11]

One by one, the rest of Kansas's five-man delegation in the House of Representatives arose to support the Skubitz amendment. Representative William R. Roy of the Second Kansas Congressional District cited a letter Governor Docking addressed to Senator Pastore, JCAE Chairman, just two days before, in which the governor had pleaded for a delay in the AEC's plans "until further scientific research can be conducted . . . to determine the safety of the project."

In the letter, Docking referred to the "bulldozer" tactics of the AEC which he said did not promote confidence in the credibility of the agency's promise to abandon the repository if hazards develop during further research on the project's safety.

An earlier letter from Docking, dated July 7, 1971, and addressed to John A. Erlewine, AEC assistant general manager of operations, was cited by Representative Garner E. Shriver of the Fourth Congressional District, which includes Rice County. In it, Docking made no concessions to niceties in putting his cards on the table. He told Erlewine:

> The final statement as prepared by the AEC offers no scientific proof of the safety of the proposed Lyons project. It offers only pledges to have faith in the AEC. Our experiences with the officials of the AEC in the past few months have given us ample reasons not to have faith in the AEC.
>
> You are ignoring the wishes of a great many Kansans when you propose—as you do in this final statement—to continue to press for construction of the repository without first conducting further tests.

When the Skubitz amendment was put to a vote, it was defeated 206 to 162. Governor Docking asked Senator Pastore to help pass the amendment in the Senate. "If the Congress approves the AEC project," the Governor said in his letter to Pastore, "Kansas Attorney General Vern Miller is making preparations to file a lawsuit to halt the project. I will support him in legal action."

When the AEC appropriation bill came up in the Senate on July 20, 1971, Senators Robert Dole and James B. Pearson of Kansas offered an amendment to restrict the AEC from starting construction on the repository for three years but allowing the agency to proceed with the demonstration project in the abandoned Carey mine under Lyons. The agency was permitted to take an option to purchase the mine acreage and the additional land, but not to execute the final purchase. This compromise was not opposed and the final appropriation bill, as amended, was passed 90 to 3. The first National Repository for Radioactive Wastes had been launched, first with assurances that it was perfectly safe and then with reassurance that if it turned out to be an environmental hazard, the project would be stopped, the site abandoned, and the wastes taken somewhere else. Where, no one could say.

There was one feature of the demonstration project that puzzled the Kansas geologists. Only 1800 feet away from the mine where the test wastes were to be buried were the active workings of the American Salt Company. Could high-level radioactive wastes penetrate that barrier once their containers had been breached? If so, salt miners in the adjacent diggings would be the first to know. The proximity of active mine workings plus evidence of some leakage in the abandoned-mine demonstration site cooled the AEC's fervor for the project. The agency's wheels began to spin in indecision.

By the end of 1971, the Nixon Administration had decided not to press for the Kansas demonstration and by the spring of 1972 the agency was considering alternate storage ideas. The new AEC chairman, James R. Schlesinger, told me in an interview that he thought the Lyons site was fading out of the picture, although burial of wastes in salt formations elsewhere was still being studied. So were alternative methods, he said, including the exotic proposal of the National Aeronautics and Space Administration to rocket loads of long-lived radioactive wastes into the sun.

In May 1972, the AEC announced a new scheme for storing solidified high-energy wastes of the nuclear-power industry on an interim basis—for a few centuries. The proposal was to build storage dumps above ground—"engineered surface facilities," the agency called them. Congressional funding for the dumps would be sought in the 1975 fiscal year and the dumps would be ready by 1979 or 1980.

Meanwhile, studies of underground storage in salt or some other formation would be continued in Kansas and elsewhere, and a pilot plant would be built to test the method. In the event of leakage, the AEC assured, wastes deposited underground at the pilot plant would be recoverable. The agency said that by the mid-

1980s it would be in a position to determine ultimately which way to go in long-term, radioactive-waste storage. Thus, the waste disposal-controversy was muted for the time being and it appeared that a final solution of the problem was being passed on to posterity.

7
Plowshare

In a nuclear explosion, energy is released in a millionth of a second in the form of kinetic energy, heat, and neutron and gamma radiation. Material at the center is heated to tens of millions of degrees—temperatures whose analogue in nature is in the interior of stars like the sun. The detonation forms a great bubble of rapidly expanding gas. Its energy is transferred to the material around it as a powerful shock wave moving outward. If the explosion is set off in an underground well, the surrounding rock will be cracked, crushed, and melted by the shock wave and heat, creating a large, flask-shaped cavity.

This effect has opened up a new dimension in man's ability to mold his environment. It provides him with a mighty shovel to shape the surface of the earth for his needs: to dig canals, tunnels, harbors, and roadbeds—literally, to move mountains—the peaceful uses of atomic energy designated as Project Plowshare.

If the bomb has not been buried so deeply that its blast effects are totally contained underground, the surface of the ground rises like an ocean swell to form a mound, which grows until it breaks apart. Underlying material propelled by expanding gases is hurled skyward. Some of it falls back into the cavity, some falls outside, and the lighter particles are projected high into the atmosphere, along with radioactive gases, to drift on the wind.

Concave slopes are formed. A crater appears, like a crater on the moon. If a series of nuclear bombs was buried in a row and detonated, the line of craters would form a canal.

At a greater depth, 1000 feet or more below the surface, the bomb explosion has a different result. While the surface may heave and shake, the energy and radiation of the explosion are contained within the cavity formed by the vaporization of surrounding rock. Initially, the cavity has a hemispheric shape. Then the roof caves in, leaving an underground chimney which extends upward from the cavity like the neck of a flask.

Contained nuclear explosions can be exploited in several ways. They can be used to "stimulate" the flow of gas or oil where the hydrocarbons are trapped in rocks of low permeability. The gas or oil simply flows into the cavity, where they can be stored indefinitely until they are brought to the surface through the well in which the bomb was buried or through another well drilled down into the cavity. This technique of underground blasting can be used also to create a storage chamber either for underground water or for natural gas piped into a region during warm weather from distant gas fields.

The advantage of using nuclear explosives instead of chemical ones for cratering or for deep-underground engineering is economic. In 1964 the AEC's Plowshare group estimated that compared to chemical explosives costing $100 a ton, nuclear explosives in 2-megaton packages would cost only 30 cents a ton, equivalent energy. Also, the nuclear devices were cheaper to emplace because of smaller bulk. There seemed to be bargains in nuclear bombs.

But the environmental drawbacks of nuclear blasting are formidable. The energy they release generates a seismic wave in the crust, that like an earthquake, can be detected halfway around the world. Besides the earth-shaking damage the blast causes buildings near the blast site, there is constant anxiety that an underground nuclear explosion might set off an earthquake.

The major environmental threat is the release of radiation produced by the explosion of the bomb. Critics of Plowshare have estimated that the cratering experiments conducted so far by the AEC have released into the atmosphere as much radioactivity as a 20-kiloton bomb exploded above the earth's surface. While the radioactivity for the most part is contained in the deep-underground shots, some escapes through cracks or fissures produced by the force of the explosion and reaches the surface. The AEC contends that with adequate geological inspection and proper implanting of the explosive, venting does not occur. Nevertheless, radionuclides have reached the atmosphere from past experimental shots, both peaceful and military in design, and, the agency's guarantees notwithstanding, are likely to do so from any future ones.

Gas or oil flowing into the bomb cavity picks up radioactive isotopes from the explosion. The most insidious of these is tritium, the radioactive isotope of hydrogen, which tends to exchange places with nonradioactive hydrogen in the hydrocarbons. The gas or oil would thus become a source of radioactive pollution wherever it is distributed—a consideration which has thus far prevented nuclear-stimulated gas from being distributed commercially.

Suez to Panama

The notion of using nuclear bombs to dig canals arose in the Suez Canal crisis of 1956 and was laid to rest, temporarily, in the Panama Canal study at the end of 1970. It may have been an idea born before its time, like the Greek mathematician Heron's steam engine.

Following the seizure of the Suez Canal by the Nasser regime in Egypt, a group of scientists met secretly at the Lawrence Radiation Laboratory in February 1957 to consider the use of nuclear explosives for excavating a sea-level canal across the Sinai Peninsula, linking the Mediterranean Sea and the Gulf of Aqaba to provide an alternate transit to the Indian Ocean. The meeting was called by Harold Brown, a physicist at the laboratory and later Secretary of the Air Force. At that time, the prospect of closing the Suez Canal appeared, at least to American scientists, a more serious threat to world commerce than the actuality proved to be. The canal was reopened in April 1957 and was closed again in June 1967 as a result of the Six-Day War between Israel and the Arab states.

Attending the Brown meeting were representatives of the

Los Alamos Scientific Laboratory, the Sandia Corporation (an AEC contractor), and the Lawrence Laboratory staff.[1] The conferees considered a variety of applications for nuclear explosives: moving earth to expose ore for open-pit mining; building water-storage basins; digging canals and harbors; creating large, underground caverns for storing gas or water; releasing gas or oil bound up in impermeable rock formations; and boring tunnels through mountains.

So enthusiastic was the response of the conferees in these discussions that a second conference, this time open to the public, was held by the AEC in May 1958 in San Francisco. By then, the petroleum industry had become fascinated with the possibilities of gas and oil stimulation by nuclear explosives, and the Continental Oil Company had begun a study of this production technique.

Brown, Ernest O. Lawrence, and Edward Teller had formed a group to explore the whole range of underground and surface engineering with nuclear explosions. In September 1957, the group had exploded a 1.7-kiloton nuclear shot coded "Rainier," deep underground at the Nevada Test Site, to confirm theories of containment and cavity formation. By the following spring, the group was able to base prospects for the new technology on limited but sound experience. Rainier and later shots provided basic data for determining the depths at which the shot should be fully contained and the size of the cavity. In the fall of 1958, the new Plowshare technology was explored further at the Second International Conference on the Peaceful Uses of Atomic Energy in Geneva, Switzerland. Delegates from a dozen nations discussed large-scale earth-moving, underground-water storage and the recharging of aquifers, the shattering of shale to release oil, the recovery of oil from tar sands, the liberation of volcanic (geothermal) heat to run turbines for electric power, and the leaching of copper ore.

With the enthusiastic support of the Joint Committee, the AEC began to promote Plowshare as the answer to just about every major earth-moving problem man could think of. One prospect after another was considered, and the evolution of each one followed a pattern which because of environmental problems has turned out to be inescapable. After its initial proposal, each project would be studied with high hopes. Extravagant claims would be made. But in the cold light of actual application, its prospects would begin to wither. The difficulties would offset the imagined advantages.

Alaska to Australia

In the five years between 1957 and 1962, ideas went up and sputtered out like roman candles. There was Project Chariot, for example, which called for the excavation of a harbor and ship-turning basin on the northwest coast of Alaska near Point Hope. Technologically, like all such schemes, it was feasible, but, like all of them, it posed the same question: Who needed it? The population of the region did not support it; they were more concerned about the possible ill effects of the nuclear explosions than the benefits the harbor might bring. There was Project Carryall, which would blast a pass through the Bristol Mountains near Amboy, California, for the Atchison, Topeka, and Santa Fe Railroad and the Interstate Highway 40. The project was declared technically feasible by the California State Division of Highways, the railroad, and the AEC. But when the engineers had estimated the actual costs and the schedule for the job, its sponsors developed cold feet. It would take too long and cost too much to justify the use of nuclear bombs for such a project. Another abort was the Tennessee-Tombigbee Water Project. Nuclear explosives would be used to blast a canal connecting two river systems, one in west central Alabama and the other in northeastern Mississippi. The scheme was dropped when the AEC concluded that the region was too densely populated for the safe use of nuclear explosives.

Was there a region that wasn't too heavily populated for the application of this technology and where it could be justified on the grounds of cost and usefulness? The search went on. There was remote Cape Kerauden in northwestern Australia where a harbor was needed so that ships could haul away the iron ore produced from a coastal deposit to an important customer, Japan.

The Cape Kerauden project contemplated the detonation of five 200-kiloton bombs emplaced 1100 feet apart at a depth of 800 feet. The explosion was expected to scoop out an elongated crater 1300 to 1600 feet in width and 6000 feet long, with a depth of 200 to 400 feet. Flooding the crater would create a harbor for the cargo ships. The AEC's partners in the enterprise were the Sentinel Mining Company, the National Bulk Carriers Corporation, and the Australian government.

Suffering the same fate as the earlier enthusiasms, this project fizzled when the partners and their experts got down to the details of carrying it out and examining its prospects. The AEC reported that the Sentinel Mining Company had re-evaluated its opportunities in the mining and marketing of the iron ore and desired to limit its participation in the project. It was understood by AEC experts that the re-evaluation had taken into account some concern among Japanese industrialists that the ore might become contaminated with radioactive isotopes from the blasting, or that the ships might pick up radioactivity—or merely that customers in Japan would think so. Thus, the project offered too great a gamble.

In any event, there was a realization that, so far as Japan was concerned, the importation of bulk ore from a region which might be contaminated by radioactivity had undesirable possibilities. The Japanese public was not ready to accept the products of atomic-bomb technology. Nor, as it has turned out, is any other public.

The Nuclear Canal

Of all the Plowshare schemes, a nuclear canal in Central America—the Panatomic Canal—was the most grandiose and seemingly the most feasible. One of the marvels of the world when it was completed in 1914, the Panama Canal became rapidly outmoded by advances in maritime technology that were able to realize economies of scale, which meant bigger and bigger ships. By 1970, for example, approximately 1300 ships afloat, under construction or on order were too big to enter the locks of the canal and another 1700 could not pass through them when fully loaded.[2] The lock chambers are 110 feet wide and 1000 feet long—dimensions which limit canal transit to ships of not more than 65,000 dead-weight tons (weight of the ship's cargo, fuel, and stores in long tons when fully laden, but excluding the weight of the ship itself). In the last third of the century, a canal was needed that would pass ships of up to 250,000 dead weight tons. Moreover, it became apparent that the rise in world commerce was producing more traffic than the canal could bear. The number of ships that can transit the canal is limited by the supply of fresh water to raise and lower ships in the locks some 85 feet between sea level and the level of Gatun Lake. As of 1970, the limit was 18,000 transits a year, according to the Panama Canal Company, and transits in that year totaled 15,000. The company was working on a $100-million-improvement program that would boost transit capacity to 26,800 a year.[3] But this would not satisfy expected traffic demand of more than 35,000 transits annually by the year 2000. The answer has long seemed to be a sea-level canal.

At the beginning of the century, the original American plan for a sea-level canal had to be abandoned because of the enormous cost of moving enough rock, especially through the Continental Divide, to bring the cut down to sea level. The United States Congress then approved the cheaper alternative—proposed by John F. Stevens and carried out by Colonel George W. Goethals—a lock canal.

The advent of nuclear Plowshare technology in the atomic age suggested the possibility of using nuclear explosives to cut the cost of building a sea-level canal which would pass the largest ships afloat. So the idea of a nuclear canal, which had been proposed at every Plowshare conference and review since 1957, took hold.

The Panatomic Canal began to assume realization in 1962, when the Secretary of the Army formed a Technical Steering Committee to review a number of previous canal studies and develop a new plan for a sea-level waterway. The technical group examined thirty possible routes which had been identified as possibilities in 1947 and picked four for detailed study.

Three of these were considered potential nuclear routes: Route 8 on the Nicaragua-Costa Rica border; Route 17 in eastern Panama (the Darien); and Route 25 in northwest Colombia. A fourth alternative was Route 14, a sea-level conversion of the existing canal by conventional methods. By 1964, studies by the AEC and the United States Army Corps of Engineers showed that on the basis of superficial calculations, at least, nuclear excavation would be cheaper than conventional canal-digging with chemical explosives and mechanical earth-moving equipment plus men with picks and shovels.

In the autumn of 1964, Congress authorized funds for a new study of the four routes, including field surveys of the feasibility and environmental impact of nuclear excavation on the Darien and Colombian routes. Meanwhile, political events in Panama, culminating in the anti-American riots of December 1964, persuaded President Lyndon B. Johnson to make two decisions. He called for construction of a sea-level canal in Isthmian America as a national goal and offered to negotiate a new treaty with Panama where the sovereignty and economic issues of the 1903 Hay–Bunau-Varilla Treaty had become the foci of intense anti-Americanism.

The following April, the President appointed a new Atlantic and Pacific Interoceanic Canal Study Commission to make the study authorized by Congress. The four routes were expanded to five in 1966, to include one just outside the Canal Zone for conventional construction of a sea-level canal.

The new study took five years and cost $22.1 million, of which $17.5 million was spent on detailed geological, biological, radiological, meteorological, and ecological assessments of the effects of nuclear explosions on the environment of the nuclear-canal routes as well as on the engineering feasibility of the technique. This was the first comprehensive investigation of the environmental impact of nuclear excavation on a large scale. In many respects, it appeared to be a model of careful, responsible work. The faunal and floral studies of the Darien and Colombian route environments contributed new knowledge to the natural history of these areas, which also were accurately mapped and sounded geologically for the first time.

The study was managed by the AEC as contractor to the United States Army Corps of Engineers. It involved experts from the Public Health Service, the Geological Survey, the Weather Bureau, and several institutes and universities acting as subcontractors. As the study proceeded, it was expected by the Study Commission that the AEC would press forward rapidly with the development of nuclear explosives, so that the effectiveness of the nuclear technique could be evaluated in the commission's final report.

Panama was the key to the realization of the Plowshare cratering concept. As Gerald Johnson, associate director of the Lawrence Radiation Laboratory told the Joint Committee early in 1965: ". . . the possibility of digging a sea-level canal across the American Isthmus is extremely important to the Plowshare program." [4] Senator Pastore, serving as Joint Committee chairman, remarked that he was willing to double, treble, or quadruple the Plowshare appropriation "if we are going to use nuclear devices for the building of a canal. If not, I would like to take a second look at the size of the budget."

But the feasibility study in the jungles of Central America and the development of nuclear explosives at Livermore, California, tended to be mutually defeating, inasmuch as one depended on the other. On one hand, the Plowshare appropriation for developing the explosives depended on the Commission's recommendation of feasibility of nuclear excavation; on the other, the feasibility recommendation depended on the development of the explosives. The Plowshare appropriation was never sufficient to enable the AEC laboratories to develop explosives during the life of the study to the point where the Study Commission could make a feasibility recommendation. A complicated array of political problems and scientific doubts unquestionably retarded the explosive-development program, but the principal inhibition was the unresolved question of environmental impact on the routes. Until this was answered, presumably by the study, it seemed fruitless to escalate the cratering test program with its attendant risk of radiation contamination and seismic disturbances.

The canal study's field investigators began their survey in the Darien by hacking a *"trocha"* or trail through the jungle to mark the line of Route 17. This route was considered quite feasible for nuclear excavation because of its remoteness from population centers. Panama City was 110 miles to the northwest and Medellín, Colombia, about 175 miles to the southeast. The only community affected was the provincial capital of La Palma with a population of 1500, on the Pacific side. The town was an aggregate of ramshackle frame buildings and *bohios* (thatched Indian huts), with a stone church, three cantina-style saloons, a cluster of stores, and a government office.

Provincial capital or not, La Palma would have to be evacuated for nuclear excavation, along with all the Cuna and Choco Indian settlements on the mainland and coastal islands in an area of 6500 square miles. About 22,000 men, women, and children would have to leave their homes for at least three years and resettle elsewhere.

Plowshare vs People

Except for the scientists in the United States who were critical of Plowshare in Panama, few Americans exhibited concern about the use of nuclear explosives to dig a canal in eastern Panama, northwest Colombia, or any other distant land. The nuclear-canal routes lay in the fastness of mountains, jungle, and swamp. Not many Americans had ever heard of the people who lived there—the Cuna Indians, who are short, wide-shouldered, thin-hipped people, many of whom are albinos. Their facial cast, with the prominent, straight nose, resembles strikingly those massive stone heads on Easter Island. Nor were the neighbors of the Cunas, the taller, darker, and more graceful Choco Indians, any more widely known in the world beyond the jungle.

About 26,500 Cuna Indians live in Panama, 25,000 of them on islands off the Caribbean coast (the Archipelago de las Mulatas) and on the northeastern coast of Panama. This region, called the Comarca de San Blas, is a special territory of the Republic of Panama. Inland, an estimated 1500 Cunas live along the rivers of the Pacific slopes of Darien, across the Isthmus from the main body of the population. The principal economic activity is fishing, but the inland Cunas also practice "slash-and-burn" agriculture—slashing the trees and burning them to cultivate the land and then moving on to clear a new site by this method when the soil at the old one is exhausted.

Descendants of, or related to, a pre-Colombian group, the Cuevas, who came from South America, the modern Cunas have a life style which has hardly changed in its basic design since the sixteenth century, except for a few conveniences, including modern medicine, sewing machines, outboard motors for some of their dugout canoes, and Coca-Cola. At the time of Balboa, the ancestors of these people dominated this region with a population of 300,000 or more. Enslavement under the Spanish Conquest, measles, and famine reduced their numbers, but could not wipe them out. Their chiefs are resolved not to give way before the new threat to their habitat—the nuclear-explosive device.

In the shaded tropical forests of espinos, laurel, and mahogany, adorned by huge morning-glories and the blood-red passion flower, time runs a different course. The Cunas speak of the coming of the Spaniards as if it happened yesterday. Tribal memory recalls Balboa as a good man, unjustly put to death by the ambitious Pedrarias. Though Balboa was beheaded at Acla in 1519, the execution is still referred to by Cunas as though it was a recent crime, and there are members of the tribe who can show you where it happened.

There is nothing primitive about these people, who live in harmony with the environment. Manning hollowed-out espave logs as canoes, most with paddles, some with outboard motors, the Indians have an efficient jungle-river and coastal transportation system. On the San Blas Islands (from which the population also would have to be evacuated in the event of nuclear cratering on Route 17) the women do a thriving business selling colorful *molas* to tourists and merchants from Panama City. Used for framing or as a pillow cover, the *mola* is a rectangle of layers of different colored cloth, cut and sewn to form a laminated multicolored pattern. *Mola* manufacture has become an extensive home industry in the islands.

In the Cuna scheme of values, there is no greater crime than one people dispossessing another. That is what evacuation means to them—dispossession. Persuading these people to get out of the way of a nuclear canal would appear to be a formidable, if not insoluble, problem in human relations. It is doubtful that they could even be bribed to move out.

On the Darien Route 17, the *trocha* runs northeasterly from the Hacienda Santa Fe, a large farm, to the Chucunaque River. The trail itself is relatively free of insects. Mosquitoes breed high above the ground in rainwater caught in the cup-shaped leaves of the bromelia vine, which climbs to the treetop by entwining itself around the trunk and branches to reach the sunlight. The mosquitoes carry the yellow-fever virus that infects a large population of howler monkeys that live in the treetops some 30 to 40 feet above the forest floor. As long as the monkeys and mosquitoes remain in their ecological niche high above the ground, man is safe. But cutting down these trees without spraying the vines first brings the yellow-fever vector to the ground.

The Isthmian section of Route 17 is 49 miles long (an additional 30 miles of approach channels in the Gulf of San Miguel on the Pacific side and Caledonia Bay on the Atlantic would be required). Nuclear explosives would be used in making the cut through the Sasardi Pass of the Divide and the Río Morti region on the Atlantic side and through the Pacific Hills on the Pacific side for a total of 29 miles. A 20-mile section in the Chucunaque Valley, lying between the hilly areas, would be excavated by conventional methods in order to assure flattened slopes and to avoid landslides in the weak clay shale of the valley.

It was estimated that nuclear excavation would require 250 explosives with a total yield of 120 megatons. The charges would be detonated in thirty groups, with yields per group ranging from 800 kilotons to 11 megatons. In order to avoid air blast and ground motion that might damage even distant population centers, sections of the route would be blown up at widely spaced intervals. Two adjacent sections would not be mined and blown in the same sequence to avoid damaging explosives or their emplacement holes. The channel would thus be excavated in two separate sequences, or "passes," each consisting of a series of alternate sections. About half the channel would be excavated in the first "pass." There would then be a delay, possibly of a year or more, until radioactivity had decayed to safe levels. Then holes for the second "pass" would be drilled and the shots would be fired in the second sequence to complete the channel.

Sequential detonation had been tested on the Upper Missouri River near the Great Fort Peck dam, Montana, where detonations of rows of chemical high explosives had blasted out a navigation channel 1300 feet long, 130 feet wide, and 17 feet deep. Compared to the plan for Route 17, however, the Fort Peck experiment was a pygmy effort. The first pass of detonations through the Divide and Pacific Hills would take about eighteen months to execute. A sixteen-month interval would follow, during which the holes for the second pass would be drilled and the nuclear bombs gently lowered into them from oil-well-type drilling masts.

The study estimated that Route 17 would cost $3 billion to build and that the job could be done in sixteen years.

The Colombian Alternative

Route 25 lay entirely within Colombia. The Pacific terminus was Humboldt Bay, 200 miles southeast of Panama City. The track crossed a narrow coastal strip and ran eastward about 10 miles through the Choco highlands, which form the Continental

Divide, with a peak elevation of 950 feet. It then swung to the northeast, crossing the Upper Truando Valley and the Saltos Highlands and running parallel to the Truando River to its confluence with the Atrato River. There it passed through a 50-mile reach of the Atrato lowlands, mostly swamp, and entered the Caribbean Sea at Candelaria Bay in the Gulf of Urabá. Including 5 miles of approach channels, the route was 103 miles long.

Nuclear blasting was feasible in a 20-mile reach from the Pacific Coast through the Divide, the Upper Truando Valley, and the Saltos Highlands. It would require 150 explosives detonated in twenty-one groups, totaling 120 megatons. As on Route 17, the shots would be fired in two passes, the first taking eight months and the second six months, with an interval of eighteen months. The largest detonation would be 13 megatons. The cost estimate for Route 25 was $2.1 billion and the estimated time required was thirteen years.

The cost estimates by the Canal Study Commission for both routes were considerably higher than those made in 1964 when President Johnson called for the sea-level canal. On the basis of the earlier estimates, the Plowshare group had pointed to substantial savings in the use of nuclear explosives compared to chemical ones. The cost of chemical explosives was constant at $100 a ton, regardless of the number of tons, while the per-ton cost of nuclear bombs dropped as the scale of energy rose. A 10-kiloton explosive might cost $350,000, or $35 a ton, while a 2-million-ton charge, with 200 times more energy, would cost only $600,000 or, as noted earlier, only 30 cents a ton.[5]

Made in 1964, the optimistic predictions turned out to be chimerical in the light of hard analysis several years later. But they served the purpose of stimulating Congressional and White House interest in the technology. In 1964, for example, the United States Army Corps of Engineers' Nuclear Cratering Group had estimated that Route 17 could be built with 170 megatons for $770 million. Six years later, after surveying the topography and geology of the route, the Study Commission raised the estimate to $3 billion, with only 120 megatons. In 1964, the Engineers had estimated that Route 25 could be excavated with nuclear charges for only $1.2 billion. Following detailed investigation, the Study Commission in 1970 pegged the cost at $2.1 billion. There were other comeuppances which began to make the Panatomic Canal look considerably less feasible after study. Even after twelve years of experimentation, the AEC admitted in 1970 that it could not guarantee the effects of yields in the megaton range. The question of radiation emissions in the light of both the Partial Nuclear Test Ban Treaty of 1963 and the Nuclear Non-Proliferation Treaty of 1969 worried the diplomatic community.

There was, of course, a long record of experimentation behind the Panatomic Canal scheme. In theory, at least, the Plowshare technologists in the AEC and in the Army knew it would work. What they did not know was the full range of environmental effects, and that was to be the problem that put the whole plan on the shelf.

The Nevada Test Shots

During the United States moratorium on testing from October 1958 to September 1961, six cratering and ditching experiments using chemical experiments had been carried out in Nevada. Four additional series of tests were made during 1960 and 1961 to determine the effects of various depths of burial of the explosives. Then, when the moratorium ended, Plowshare returned to nuclear testing. The two landmark nuclear tests on which Panatomic Canal technology was largely predicated were Danny Boy, a 0.43-kiloton shot on March 5, 1962, and Sedan, a 100-kiloton shot on July 6, 1962. The small Danny Boy scooped out a 220-foot-wide crater, 60 feet deep, in basalt, a volcanic rock, and Sedan excavated a crater 1280 feet in diameter and 320 feet deep, in looser Nevada alluvium. The tests showed that cratering technology worked—that if you buried enough bombs in the right places, you could blast a trough in the surface of the earth anywhere.

There followed, then, a series of tests designed to demonstrate the differential effects of blasting in different kinds of rock. After the 1963 Test Ban Treaty, a shot coded Sulky was set off in basalt at a depth of 90 feet on December 18, 1964. Research men charted the dispersion of air-borne radionuclides from it. In another test coded Palanquin on April 14, 1965, the experimenters set off 4 kilotons in rhyolite, another type of volcanic rock, and again, the dispersion of radionuclides was chartered. The agency was trying to devise "clean" explosives and develop a blasting technique that would keep most of the radioactive debris in the crater. That was a critical requirement for Panama, in the light of the Test Ban Treaty forbidding the dispersion of radionuclides from nuclear events across borders.

Radiation Exposures

Was it possible to keep radionuclides confined from a subsurface detonation which inevitably vented to the atmosphere? Lawrence Radiation Laboratory researchers expected that the following would be the pattern:

At least 50 per cent of the total radioactivity produced during the nuclear excavation of a sea-level canal should be contained in the fallback deposited in the channel.

Another 40 per cent would be deposited near the channel in material thrown out of the crater and in dust from the base surge —the near-ground-level smoke ring that spreads outward from a nuclear explosion.

That left 10 per cent for wide distribution as fallout or rainout.[6]

The Battelle Memorial Institute, which managed the radiological safety studies for the AEC during the Canal Study, maintained that beyond close-in areas, "potential external radiation doses will be far below the lethal levels and well below the levels known to produce obvious clinical symptoms in man." [7]

But what about the levels which do not produce "obvious" clinical symptoms in man—but which are now suspected as the cause of statistical increases in the incidence of leukemia in children as well as this and other forms of cancer in the general population?

As far as the Canal Study was concerned, low-level radiation effects were not even considered. But the fact that radiation would be injected into the biosphere was calculated. The study noted that human populations living in the fallout area "may be exposed to some external radiation and to internal radiation through the food chain."

The extent of fallout would depend to some extent on the height of the main cloud above the doughnut-shaped base surge. The higher the cloud, the more widely would the fallout be scattered.[8]

Using the 1962 Sedan crater test as a scale, two investigators of the Air Resources Laboratories, Environmental Science Services Administration, made a calculation. The main cloud from the Sedan test rose to an altitude of 12,000 feet and the base surge to 4000 feet. The investigators, Gilbert J. Ferber and Robert J. List, estimated that cloud heights from canal explosive yields might range upward from 20,000 to 40,000 feet, with the base of the stratosphere at 55,000 feet as the upper limit.[9] They calculated also that the whole-body gamma-radiation dose in close areas would exceed 3.9 roentgen, which was the standard set as the maximum permissible lifetime dose to the population near the Nevada Test Site when the United States was testing nuclear bombs in the atmosphere.

Army engineers continued their chemical-cratering explosions while the AEC experimented with nuclear devices during 1965, 1966, and 1967. On January 26, 1968, the third big Plowshare cratering experiment (called Cabriolet), the detonation of a 2.5-kiloton bomb, was executed in rhyolite at the Nevada Test Site. It produced a 360-foot crater, 120 feet deep. A row of five 1.1-kiloton nuclear bombs buried 150 feet apart at a depth of 135 feet was then detonated March 12, 1968, in a test called Buggy I. Their explosion produced a crater 850 feet long, 250 feet wide, and 65 feet deep. Next, on December 8, 1968, the AEC exploded a 35-kiloton charge in tuff, a common porous rock at the Test Site. The experiment, called Schooner, dug a crater 800 feet in diameter and 270 feet deep.

Still, the Division of Peaceful Nuclear Explosives had not yet reached its goal of clean, ideally shaped charges. Seaborg had warned the Joint Committee in 1965 that developing adequate nuclear explosives for canal excavations would take at least five years. But at the rate the AEC's Plowshare experiments were proceeding, it did not seem likely that the explosives could be developed before the Canal Study Commission completed its five-year survey in 1970.

Radiation Pathways

In the jungles of Panama and Colombia, teams of engineers, geologists, biologists, medical ecologists, radiologists, and meteorologists toiled for three years. Drilling rigs to obtain rock cores 2000 to 2500 feet underground were hoisted into position on the slopes of the Continental Divide by military helicopters. The cores were clues to the mountain structure of the Divide. Plant ecologists and soil chemists collected specimens and samples to predict the concentrations of radionuclides that might be expected in various species as a result of uptake through the soil.

A marine bioenvironmental study was carried out to ascertain the distribution of radionuclides from the blasting in coastal waters. Commercial fish caught in the Gulf of Panama, such as shrimp and tuna, which are shipped to world markets, could concentrate radionuclides falling into the Gulf from the explosions. Ecologists studied the diets of animals and man in the regions of the San Blas, Darien, Choco, and Antioquia, which were likely to become contaminated by radioactive fallout. Agricultural methods used by the Indian and Latin populations of these areas were analyzed to trace possible paths of radionuclides through the food chains.

Fresh-water ecology studies called for analysis of every type of fresh-water fish in the diet of the Indian population living near streams in the blast areas of both Routes 17 and 25. In this investigation, three Peace Corps workers who had been working with the Choco Indians proved invaluable, for their relations with the Indians were so good that they were able to recruit Choco assistance in catching the fish to be analyzed. The biologists needed to know what the fish ate in order to trace the pathway of radionuclides that might fall or be washed by rain into rivers and streams. A paralytic drug that the Chocos make from tree bark was dropped into the water and soon fish began floating to the surface. In one day, more than 100 pounds of fish taken both by means of the drug and by Choco fish spearmen were turned over to the biologists to be frozen and shipped to the Battelle Northwest Laboratory at Richland, Washington, for dissection and analysis.

The Environmental Science Services Administration set up weather stations at the terminals of each route to chart wind patterns for fallout dispersal.

Marine resources were studied by a University of Miami group. Estimates of radioactivity dosages were made by the Oak Ridge National Laboratory. Tidal currents from the Pacific in a sea-level canal were computed by a Massachusetts Institute of Technology group. Medicoecology studies, forecasting disease vectors and health conditions on each route, were carried out by the Office of the United States Army Surgeon-General. Although it was bogged down from time to time by funding uncertainties, the environmental-impact study was the most ambitious in the history of nuclear technology.

So far as the radiation estimates were concerned, there were a number of imponderables. The chief one was the amount of radionuclides the blasting would release to the atmosphere. The AEC was working on cleaner bombs, but the estimates had to be based on older explosive experiments which had been moderately dirty.

In general, the studies "lead to the belief that biological concentrations of radioactivity at levels that would endanger humans would not be likely to result from nuclear excavation along any of the routes considered," the Commission's report concluded. That point of view had persisted all through the investigation, which entertained a highly optimistic outlook toward the biological hazards of the bombs, even though fallout was expected to run as high as 10 per cent of total radioactivity released from the explosions.

The investigators themselves seemed to adapt readily to jungle conditions. The Route 17 headquarters at Santa Fe in eastern Panama boasted three Army-style barracks buildings, included a day room, aid station, and radio-telephone service capable of reaching Panama. Because of a shortage of military helicopters, which were tied up in Vietnam, transportation between Route 17 camps and Panama City was provided by small, private airlines boasting two or three single-engine, light, and well-used aircraft. Even GI equipment was scarce. Army specialists complained that no jungle boots were available. They were being shipped to Southeast Asia.

On a general reconnaissance, a project officer gave this description of what it was like in the Colombian jungle:

> At this point in time, we were a pretty bedraggled and tired lot. The hike to the Divide was much easier [than expected] and a hard rain that lasted forty minutes cooled things off considerably. Incidentally, it does not seem to be important to attempt to protect yourself from the rain. You are so soaked through by perspiration that the rain actually feels cool and refreshing. . . .

The Opposition Gains

Events beyond the power of the engineers and scientists to assess were conspiring to chill nuclear-canal prospects. In Panama and Colombia, the probability of fierce resistance of the Cuna and Choco populations to evacuation and resettlement was a critical obstacle. The Canal Study Commission took the position that, although pioneering in such a massive nuclear-excavation project would impress the world, any attempt to proceed against the will of the local populations would mar the national image. In December 1967 two scientist critics, David R. Inglis at the AEC's Argonne National Laboratory and Carl L. Sandler, observed in the *Bulletin of the Atomic Scientists,* "Against the short-term economy of the nuclear method must be balanced our lack of full information on its use and the consequences of increased radioactivity, the need to clear the adjacent areas and to relocate the population. None of these reservations is to be taken lightly."

E. A. Martell, a senior scientist at the National Center for Atmospheric Research, reiterated estimates in the magazine *Environment* that 4 to 10 per cent of the total radioactivity from cratering shots would appear as local fallout. The estimates were based on results of the Sedan and Danny Boy Plowshare tests. Martell said:

> This substantial atmospheric contamination with radioactive isotopes of greatest biological significance to man [strontium-89 and -90, cesium-137, iodine-131, and tritium] by nuclear cratering shots poses an especially serious hazard for the nuclear canal project. Because the heights of the debris clouds from high-yield nuclear cratering shots in a complex medium are very uncertain, both the transport of the cloud debris in the variable wind systems . . . and the resulting radioactive fallout areas would be highly unpredictable.[10]

The question of whether the Limited Nuclear Test Ban Treaty would be violated remained unsettled. During authorization hearings on the study in Congress, it was reasoned that if the feasibility of nuclear-canal excavation could be demonstrated, the constraints of the treaty might be modified. Some support for this hope had appeared in the Treaty of Tlatelaco, the Latin American nuclear-free-zone agreement, in which fifteen Central and South American countries, including Panama and Colombia, agreed to ban nuclear weapons from their territories but made a special provision for employing nuclear explosives for peaceful purposes. Also, the Nuclear Non-Proliferation Treaty, in Article V, indicated the possibility of international agreement, subject to modification of the Limited Nuclear Test Ban Treaty, to allow nuclear-excavation projects under appropriate international observation.

The Panamanian Standoff

In spite of these developments, the Study Commission by 1970 realized that "prospective host country opposition to nuclear-canal excavation is probably as great if not greater than estimated in 1964." [11]

The opposition was in Panama, where it was rooted in pride of sovereignty, a pride which rejected extension of American dominion as intolerable. This native hostility toward the Colossus of the North had been growing for generations. It was fed by disputes over wage differentials between American and Panamanian Canal employees; by the obvious economic advantages of Zonians (Americans living in the Canal Zone) compared with the living standards of the Latins and Jamaicans employed in the Zone and on the Canal. These feelings of political and economic subjugation were exploited by Cuban and Chinese agents and by nationalists as well, so that disputes over such symbolic representations of the underlying conflict as the flying of United States or Panamanian flags in the Zone erupted into the anti-American riots of 1959 and 1964.

The legal basis of political hostility was the Canal treaty negotiated in 1903 by Theodore Roosevelt's Secretary of State, John Hay, and the French adventurer, Philippe Bunau-Varilla, representing Panama. The treaty was designed to induce the United States to take over the assets of the defunct French New Canal Company which had been looking desperately for a buyer since it went bankrupt in 1899. In this design, the treaty was successful. It allowed the French to salvage part of the investment in their ill-fated attempts to build an Isthmian canal which had been inaugurated by Count Ferdinand de Lesseps, builder of the Suez Canal. But the effect of the treaty on the fledgling Republic of Panama was to subject it to a form of international peonage as a protectorate of the United States. No matter how humanely the peon was fed, clothed, and housed, he was still a peon. Politically, the status was outrageous in the view of Panamanians who believed their country had a right to seek its own destiny.

The Hay–Bunau-Varilla Treaty was drafted in an era when the United States, having flexed its muscles by dislodging Spain from Cuba and its Pacific possessions, discovered it had vital interests in the Caribbean. The Spanish-American War had dramatized the strategic value of an Isthmian canal as a military as well as a commercial link between the Atlantic and Pacific coasts of the United States.

The treaty itself was the result of a series of maneuvers in which the United States gained access to Panama—and exclusive access. In the preliminary move to get canal rights, Hay persuaded Great Britain to agree to unilateral United States control over an Isthmian canal. This was formalized in the Hay-Pauncefote Treaties of 1901–1902. Then Hay began negotiations with Colombia for canal rights in Panama—which was then a Colombian province, albeit an unhappy one. A treaty was hammered out but nationalist sentiment in the Colombian senate persuaded that body to reject it. Events then became melodramatic, as they frequently do in Central America on short notice. Panama revolted and set up a provisional government. United States Navy ships discouraged Colombia from landing troops in Panama to quell the revolt. Three days after Panama announced its independence from Colombia on November 3, 1903, Washington recognized the revolutionary junta and the canal treaty was promptly negotiated and approved. Stockholders in the French canal company had been saved.

The 1903 treaty created the ten-mile wide Panama Canal Zone extending across the Isthmus and placed it under the sovereign control of the United States in perpetuity. It gave the United States a canal-construction monopoly, the right to police Panama City on the Pacific side and Colón on the Atlantic side if Panama could not maintain order. It gave the United States the right to station troops and build fortifications in the Zone. Panama did not even have the power to tax the Panama Canal Company, a United States government corporation.

In return, Panama received payment of $10 million in gold and an annuity of $250,000 for the canal concession. After the canal opened in 1914, Panamanians realized that the Americans had come to stay. An American enclave arose on the American side of the Canal Zone fence adjacent to Panama City. Except for its architectural adaptation to the humid, tropical climate and barracks-style dwelling units, it was culturally a Midwestern American small city called Balboa, with a population of 4000. It had its own government, its own smartly uniformed police force wearing wide-brimmed straw hats, its own school system, and a commissary which evolved into a chain of supermarkets. The Zone was middle-class Americana—with its air-conditioners; refrigerators; garbage-disposal units; shiny, big cars; radio and television sets,—and it was as effectively segregated from the culture of Panama as though surrounded by a high wall. The American military establishment in the Zone—the United States Southern Command—provided housing, schooling, shopping, and entertainment for the troops and their dependents, plus a bland fare of radio and television, replete with old stateside shows, for the entire Zone population.

On the Atlantic side of the Isthmus, a community considerably smaller than Balboa but similar to it grew up around the port of Cristobal, adjacent to the old Spanish city of Colón. These United States enclaves offered a persistent and highly visible contrast between American affluence and the spectacular poverty of the teeming barrios of Panama City and Colón. And it was the symbol of the Panamanian grievance.

Latin political logic may appear to be both intricate and bizarre to North Americans. It was difficult for Zonians to understand why the Latins blamed their woes on the canal treaty, inasmuch as the Canal provided the country's largest single source of income. Even the financial and political concessions with which the 1903 treaty has been modified since 1936 have failed to satisfy national resentments. Treaties providing for eventual Panamanian control of the canal and giving the United States an option to construct a sea-level waterway were negotiated in 1967 during the administration of President Marco Aurelio Robles, regarded as pro-American. The treaties were denounced by Robles's chief opponent, Dr. Arnulfo Arias, and became a campaign issue in the election of 1968, which Arias won by a big margin, although he was ousted a few days later by a revolutionary junta. Since then, no politician has either desired or dared to defend the treaties. In the fall of 1971, the junta notified Washington that if new treaty talks did not produce an acceptable canal arrangement, Panama would seek mediation from the United Nations—no empty threat.

The Denouement

Such was the setting in which the United States considered the construction of a sea-level canal with nuclear explosives in the late 1960s. Politically, the effort was doomed from the start, not only because of unacceptable Panamanian treaty demands for control of the Canal and of the Zone, but also by the resistance of the Latin and Indian populations in the Darien. Nevertheless, the Canal Study could be rationalized by the hope that the United States could make an acceptable deal either with Panama and appease the Darien residents in some manner, or with Colombia for Route 25.

Political problems in Central America were not the only ones to be reckoned with. There were problems at home. By mid-1968, it became evident that budget constraints on the AEC would not permit the development of nuclear-excavation technology to the point where the Atlantic and Pacific Canal Study Commission could determine its feasibility. Neither Congress nor the White House was pushing a nuclear canal. Scientists were speaking out against it as a dangerous experiment. Some raised the question of adverse environmental effects of a sea-level waterway that would allow the mixing of marine fauna from the two oceans.

The revival of the radiation controversy on the hazards of bomb fallout in the United States in 1969 further cut prospects of any substantial funding of Plowshare at a level permitting the development and testing of explosives for canal digging.

As a result, by 1970, the Canal Study Commission's Technical

Associates for Geology, Slope Stability, and Foundation were "in unanimous agreement that the techniques for nuclear excavation of an interoceanic canal cannot be developed for any construction that would be planned to begin within the next ten years." [12]

The Technical Associates held that results of the relatively small-yield tests in the kiloton range which the AEC had conducted could not be adequately extrapolated to higher yield effects. The largest cratering experiment the AEC had undertaken was the 100-kiloton Sedan blast in 1962. Excavation on Routes 17 and 25 called for yields of 3 megatons per shot and for salvos up to 13 megatons. "Extension of the scaling relations now established by tests to the much higher yield explosions is too indefinite for assured design," the Technical Associates report said.

In its report to President Nixon, the Commission recommended the construction of a sea-level canal by conventional means on Route 10 in Panama, at an estimated cost of $2.88 billion. This route is just outside the present Canal Zone, about 10 miles west of the existing canal. The route is 53 miles long and would consist of 17 miles of two-lane approach channels and 36 miles of a single-lane land cut. The Commission was optimistic that a sea-level canal on Route 10 would be acceptable to Panama "under reasonable treaty conditions."

It appeared that Project Plowshare had lost the ball game with the AEC's failure to convince the Canal Study Commission's experts that the new technology was ready for big jobs. AEC Chairman Seaborg acknowledged that public concern for the environment and the Limited Nuclear Test Ban Treaty had slowed development of cratering technology, but he added: "It is our view that given the authorization and funds the problems regarding technical feasibility can be solved within a relatively short time." An explosive specifically designed for excavation, Seaborg said, had been developed in nine tests; the last one, coded FLASK, which was carried out in May 1970, showed that radioactivity had been reduced by a factor of five below previous levels.

"We believe that if for any reason a decision to construct an interoceanic, sea-level canal is delayed beyond the next several years," Seaborg added, "a nuclear-excavation technology might be available and provide a realistic option in canal construction considerations at that time." [13]

The AEC budgets for 1971 and 1972 fiscal years lent scant encouragement for any rapid future development of nuclear-cratering technology. In the jungles of Central America, Plowshare had made its bid to demonstrate a global, terra-forming technique, and lost. Outside of the AEC and the Joint Committee, there was little support for exploding nuclear devices at so shallow a depth that their radioactive debris immediately reached the biosphere.

Cratering became the left hand of Plowshare. The right hand was still promising—contained underground blasting to liberate oil and gas and to create storage cavities. This technology had the support of the oil and gas industries—powerful allies. Moreover, it was not confronted by stubborn Indians and angry Latins. But it faced another adversary that might prove overwhelming—the native American environmentalist movement.

8
Bonanza from Bombs

The Atomic Energy Commission warns frequently of the rapid depletion of oil and gas in justifying expenditures for nuclear-reactor development, which totaled $311 million in the 1971 fiscal year. Nuclear power, the AEC argues, can help this nation hoard its dwindling supplies of oil and gas. In promoting the breeder reactor on this basis, Commissioner Clarence E. Larson cited an estimate [1] that United States resources of oil would be depleted by the year 2000 and those of natural gas by 2015. Coal would probably last until 2400.[2]

Yet, the AEC proposed in 1971 to spend $85 million to develop nuclear explosives and techniques of using them deep underground to accelerate the recovery of America's oil and gas resources for the benefit of private industry. The inconsistency between what the AEC says and what the AEC does is one of the consequences of the lack of a national energy policy.

Underground engineering—the use of "contained" nuclear explosions to liberate oil and gas trapped in tight rock formations—has been the aspect of Project Plowshare of greater immediate interest to industry than cratering. For it appears to offer more immediate profits, with less apparent risk to public health and safety, than cratering technology.

However, nuclear explosions deep underground are not always contained. Between September 1961 and June 1963, radionuclides seeped to the surface in 27 out of 70 underground detonations.[3] A five-year study of 900 mice on the AEC's Nevada Test Site, where most underground testing is conducted, showed that low-level radiation hastened the processes of aging in the mice, shortened the reproductive period in the females, and reduced the number of young per litter.[4] Moreover, the likelihood that the underground explosions can set off earthquakes or inject radionuclides into ground-water channels, cited by conservationists in their protests against the use of the technology, has not been refuted effectively by the AEC.

So long as the explosions were confined to the Nevada Test Site, the underground-engineering development program was relatively free of intervention by citizens. But since the new technique of loosening up oil and gas from impermeable underground rock formations offered the hydrocarbon industry a relatively cheap way of tapping new gas and oil resources, the program did not long remain confined to the government reservation. It was quickly moved out to the gas fields of New Mexico and Colorado and the use of the technology was seriously promoted in central Pennsylvania.[5] "Nuclear stimulation," as the technique is called in the petroleum industry, looked like a real bonanza for the oil and gas industry. It was a gift from a government agency concerned, on one hand, with the conservation of oil and gas resources and eager, on the other hand, to enable the industry to pull them out of the ground as rapidly as possible.

There is a distinction in this industry between the term "resources" and the term "reserves." When oilmen talk about "reserves," they usually refer to deposits in the ground which can be recovered by methods and machinery they are presently using at a cost which will guarantee a profit. From this point of view, oil and gas reserves are like money in a checking account. As long as they last, they can be drawn upon.

The term "resources" refers to deposits which are known to exist beyond the readily recoverable reserves, but which are more difficult and costly to reach. The additional resources are not exploitable on the same economic basis as reserves. In most instances, it costs more money to extract them than the current market prices would justify.

If, however, a new technique, such as nuclear stimulation, could be widely used to loosen up tightly bound oil and gas resources and make it economic to recover them at current market prices, these less accessible deposits would become immediately exploitable. And they would add immeasurably to the wealth and financial prospects of the companies that could get them—until, of course, the resources were exhausted.

Without nuclear-stimulation technology, existing reserves show a relatively imminent depletion. By 1970 it became evident that the known reserves could not long continue to supply rising demand. A gas shortage impended, despite the existence of vast resources in the Rocky Mountain region and in the Southwest—resources which were not economical to reach by conventional means.

The ratio between reserves and resources of crude oil, for example, was indicated in 1961 by these estimates of the American Petroleum Institute and the Department of the Interior: reserves were estimated at 48 billion barrels and "ultimate reserves," or resources, at 140 to 300 billion barrels. Other "ultimate reserves" estimates ranged from 364 to 460 billion barrels.[6]

Reserves of natural gas were estimated in 1961 at 267.7 trillion cubic feet by the American Gas Association.[7] At current and anticipated consumption rates at that time, the reserves were good for less than forty years. Other estimates of the total gas supplies ranged from 510 to 2650 trillion cubic feet.[8]

The Contaminated Treasure

Largely because nuclear explosives come in smaller packages and are much cheaper than chemical ones, the use of nuclear charges deep underground pointed the way toward the recovery of hard-to-get oil and gas at a reasonable cost. But side effects had to be considered. In addition to damage from earth shaking, there was the problem of radioactivity in the gas or oil.

The effect of the underground explosion, as I have related earlier, is to create a flask-shaped cavity in the rock, into which flow the gas and oil from surrounding rock fractures. Since the radionuclides produced by nuclear fission remain in the cavity, the hydrocarbons inevitably become contaminated with them. At least a year is required to allow the radionuclides with shorter half-lives to decay before the gas or oil could be brought up for distribution. After that, the likelihood of residual contamination with the longer-lived radionuclides would always pose a marketing problem.

It appears that both the government and the hydrocarbon industry overlooked or underestimated these effects in their first flush of enthusiasm for the new technology. The first nuclear containment shot, a bomb with an explosive yield of 1700 tons of TNT, was exploded in volcanic tuff at Mercury, Nevada, in 1957. The nuclear event was hailed as the opening of a new era for the gas and oil industry.

"Here for the first time was a physically small package of an explosive having extremely high energy release in the form of heat and pressure. Perhaps a nuclear explosion could be used to stimulate petroleum production," observed J. Wade Watkins and C.C. Anderson of the United States Bureau of Mines.[9]

The Richfield Oil Corporation proposed to the AEC that a nuclear device be used to produce tar from the McMurray Tar Sands in Canada. The Bureau of Mines proposed a test in thick oil-shale deposits in the Rocky Mountains. Neither proposal advanced beyond the talking stage, but the potential benefits of the new technology had become obvious. The Bomb at last would bring a bonanza to the oil and gas industry. It would create new millionaires overnight. It would boost the stock of oil and gas companies to undreamed-of heights. Gas and oilmen thought not of one or two stimulations, but of thousands of them, all over the west, in Alaska, and wherever geologists suspected the presence of large deposits of hydrocarbons in low-permeability formations. In the deep-underground engineering mode, the nuclear scourge could become the open-sesame of a buried treasure, just beyond the economic reach of conventional extraction methods.

By 1969 the AEC reported that "containment" shot technology was "safe" to use. The agency said it had exploded two hundred nuclear devices underground in vertical drill holes since 1961 and that "results to date show that containment and venting problems are reasonably well understood."

Ten of these "events" showed what the AEC defined as "significant leakage." All but one of these were under 20-kiloton yields and one was in the range of 20 to 200 kilotons. None of the higher-yield explosions showed any "significant" radioactive leakage, the AEC reported. This meant that an undetermined number of them showed some leakage, which the AEC did not feel was significant. Three of the ten events which leaked significantly vented through ground fissures, starting within the first minute after the explosion. From these experiences, the agency advised the Joint Committee, it had been possible to refine the determination of the depth at which leakage would not occur.

The gas industry readily saw the possibilities of stimulating gas flow in the tight, impermeable gas sands and rocks of the sedimentary basin under the Rocky Mountains. By the end of the 1960s the United States Bureau of Mines had estimated that 317 trillion cubic feet of gas were contained in these clogged reservoirs, where the gas was held in the pores of the rock. Normally, some of it would flow through tiny fractures and connections between the pores to a well, but it would not move fast enough to make the sinking of a deep well into the formation economically sensible. By breaking up the rocks with a bomb, the gas could be freed. A nuclear bomb was better than a chemical one because it was not only much more powerful, but it was cheaper and more compact and could be lowered down through a smaller, less costly well.

Gasbuggy

Of the 317 trillion cubic feet of gas thought to lie in the depths of the Rocky Mountain region, about 199 trillion were located in the Green River Basin of Wyoming and 118 trillion were distributed in the Uinta Basin of Utah, the Piceance Basin of Colorado, and the San Juan Basin of New Mexico. In 1965, the El Paso Natural Gas Company submitted a proposal to the AEC and the Department of the Interior for a test of nuclear gas stimulation by a 20-

to 30-kiloton shot in the San Juan Basin of northwest New Mexico. The experiment, titled "Gasbuggy," was, at an estimated cost of $4.7 million, to open a new era for the gas industry. Or so the promoters hoped.

Preparations went forward slowly, paced by the federal budget cycle. By 1966, the engineering feasibility of the project was established. "Based on present knowledge of areas where nuclear explosives may be effective in improving gas production, more than 30,000 such shots (of the order of the one being prepared for Gasbuggy) have been projected for the Rocky Mountain Region alone," C. H. Atkinson of the Bureau of Mines and Sam Smith of El Paso Natural Gas told a 1966 regional conference of the American Institute of Mining, Metallurgical, and Petroleum Engineers in Seattle. "Should the Project Gasbuggy experiment confirm predictions made in the feasibility study, and we are optimistic that it would, the nuclear-explosive fracturing technique could become almost routine within five to eight years after the Gasbuggy detonation. In short, this could be the beginning of a vast, new enterprise for more efficient recovery of hydrocarbons." [10]

The Lawrence Radiation Laboratory was hardly less enthusiastic. At a development rate of fifty wells a year, more than 10 per cent of the nation's 1970 gas consumption could be met by gas from nuclear stimulation, laboratory personnel reported.[11]

However, a staff analysis of this prospect by the Federal Power Commission commented that it would take twenty years before that level of gas production could be met by nuclear-explosive stimulation. A level of production equal to 10 per cent of 1969 gas consumption would require the explosion of 4000 nuclear devices of 100 kilotons each (400 megatons) in 1000 wells over the twenty-year period.[12]

Radioactive Gas

After two years of preparation Gasbuggy, the first joint experiment by government and industry in deep-underground engineering with a nuclear bomb was ready. At 12:30 p.m. on December 10, 1967, a 26-kiloton nuclear explosive was set off 4240 feet below the surface of land leased by El Paso Natural Gas about fifty-five miles east of Farmington, New Mexico. The technical direction of the shot was provided by the Lawrence Radiation Laboratory at Livermore.

The ground shook, but there were no other significant seismic effects. No release of radioactivity from the explosive was vented, the AEC reported, but radioactive gas under high pressure seeped upward through the insulation of the electronic cables in the drill hole. The cables were then cut and sealed. Ground motion from the shot was felt in Dulce, in the Apache Indian Reservation fifteen miles away, and in other communities within a radius of thirty-five miles.[13]

Several hours after the detonation, workers began to drill into the cement poured to fill the hole into which the explosive had been lowered. The drill penetrated a zone of broken rock in the natural-gas-bearing sandstone formation by January 10, 1968. A chimney whose top was 3907 feet below the surface, or 333 feet above the point of detonation, had been formed.

One of the critical test objectives, of course, was to determine the radiological safety of such a shot. Surveillance by the Southwestern Radiological Health Laboratory of the United States Public Health Service, Las Vegas, confirmed that no radioactive material was released into the biosphere from the shot, nor was any injected into adjacent gas wells.

During 1969 hearings by the Subcommittee on Air and Water Pollution of the Senate Committee on Public Works, Senator Muskie, the Subcommittee chairman, observed that no reference was made in the summary of the report to radioactive contamination of the gas collecting in the shot cavity.

"It seems to me that if there was no examination of the results, that should have been stated, or, if the results were unfavorable, that should have been stated," Muskie said.[14]

The Lawrence Radiation Laboratory reported that about 4

grams of tritium had been created in the shot hole by the explosion and tritium was present in the gas.[15] Krypton-85 and argon-37 were found, but no radioactive iodine was seen, the report stated.

The concentration of tritium in the Gasbuggy well was too high to permit the gas to be distributed commercially without considerable protest, but it was released to the environment anyway when the gas was flared (burned at the well). Radiochemical analyses, along with flow rates, indicated that 1860 curies of tritium and 290 curies of krypton-85 were produced through February 3, 1969, in the first 161 million cubic feet of gas. On this basis, scientists of the Oak Ridge National Laboratory and El Paso Natural Gas predicted that a total of 2500 curies of tritium and 350 curies of krypton-85 would appear in the gas during the lifetime of the well. The scientists concluded: "The hypothetical dose equivalents to various population groups were well within the annual dose limits prescribed by the International Commission on Radiological Protection and other authorities when the single well was considered." [16]

But at this level, total doses from the thousands of wells envisaged by Plowshare promoters would create a serious hazard. The investigators urged caution in establishing permissible concentrations of man-made radioactivity in natural gas for industrial and domestic distribution. "We base our caution on the fact that nuclear stimulation of natural-gas reservoirs represents only one of the many potential sources of radiation exposure and that an extremely large population is involved." [17]

At the end of 1969, cumulative production from Gasbuggy reached 213 million cubic feet. Over a period of twenty years, total gas recovery from the well was estimated at 900 million cubic feet, a fivefold increase over the estimated production of conventional wells in the area. At the end of production tests, the rate of gas flow was 160,000 cubic feet a day—six to seven times that of the other wells in the area.

"It now appears that tritium is the radionuclide which may

cause the greatest difficulty if the stimulated gas is to be distributed in pipelines for normal industrial and home use," Richard Hamburger of the AEC told a panel on Peaceful Nuclear Explosions in Vienna in March 1970.

Since no nuclear explosions had ever been set off in a geologic setting similar to the site in the San Juan Basin, geologists were curious about the seismic effects. El Paso Natural Gas Company investigators reported that "observed surface accelerations were about twice as much as expected based on previous nuclear explosive experiments." Gasbuggy provided "a new basis for seismic evaluations of other prospective gas-reservoir stimulation experiments where the gas bearing formations outcrop within forty miles of populated areas." [18]

Could these detonations awake dormant seismic forces underground? Could they trigger earthquakes? The AEC minimized this possibility; its own test experience had given no indication of seismic effects. Yet, not long after Gasbuggy, the agency exploded a test charge in the megaton range at 4600 feet below Pahute Mesa in the Las Vegas Bombing and Gunnery Range. Within four weeks 10,000 tremors were recorded, and of the 500 for which points of origin were determined, all but 3 had originated within eight miles of the blast site.[19]

The Rise of Project Ketch

In addition to gas and oil stimulation, underground nuclear blasting offered another prospect to the gas industry—the creation of subterranean storage chambers in the cavities created by nuclear explosives. With the increased use of natural gas for space heating and industrial processes in the first half of this century, it became necessary for gas suppliers to find places where gas could be stored during low-use periods in warm weather when the source of the gas was hundreds or thousands of miles away from the customers. Otherwise, suppliers would hardly be able to meet peak demands during cold weather. The alternative to storage reservoirs was a long-distance gas-transmission system with enough capacity to handle peak demand—a solution that the industry contended would be uneconomical.

With 28.3 million residential customers of gas in 1967 and 43.3 million expected by 1980 [20], the gas-storage complex which existed in the mid-1960s would have to be greatly expanded. At the beginning of 1966, there were 308 underground gas-storage reservoirs in twenty-five states, representing an investment of $1.25 billion. Of 293 reservoirs in actual operation, 238 were in depleted gas fields and 55 were in oil and gas or oil fields, exhausted aquifers (which once had carried water), an abandoned salt mine, and an abandoned coal mine. Gas was also stored in cryogenic (supercold) tanks in liquid form at 260 degrees below zero Fahrenheit, to reduce its volume. There were four such cryogenic storage units in 1966 and six more were being constructed.[21]

Such was the gas storage picture at that time. It provided 4.3 trillion standard cubic feet of capacity. The industry said it could have used another 5 trillion if this amount could be obtained at low cost.[22] Perhaps it could be obtained economically by means of a deep-underground nuclear explosion in an impermeable rock formation, creating a subterranean storage cavern.

A proposal to test this idea was made to the AEC by the Columbia Gas System Service Corporation, a gas-distribution system, and the AEC agreed to it. "Storage facilities so developed offer the potential of being economically superior to mined cavities and natural gas facilities," the group's prospectus said. Although a "significant body of data" had been assembled on the physical effects of nuclear explosions, the prospectus continued, "it is concluded that an experiment is needed to answer basic questions. . . ." These included the capability of a nuclear bomb-created reservoir to store gas at various pressures and the economics of commercial application of this method of creating storage.[23]

The experiment was designed by the AEC's San Francisco Operations Office, the Lawrence Radiation Laboratory, and the Bureau of Mines working with the Columbia Gas System. In the peculiar, nautical nomenclature adopted by the AEC for this series of tests, it became known as "Project Ketch," and called for detonating a 24-kiloton nuclear explosive at a depth of 3300 feet in a thick, impermeable shale formation in central Pennsylvania. The scene of the proposed detonation was the largest forest in the state, the Sproul State Forest in Clinton County, at a site about twelve miles southwest of the town of Renovo.

The Lawrence Radiation Laboratory calculated that the detonation would create a chimney 300 feet high and 180 feet wide, providing storage for 465 million cubic feet of gas at a pressure of 1200 pounds per square inch. Fracturing of the surrounding rock was not expected to extend beyond 650 feet above the point of the explosion. The void space thus created would contain radionuclides, but the chimney could be purged by gas to remove most of them and the gas could be filtered and treated at the wellhead to remove additional contamination. None of these procedures, however, would remove the radioisotope tritium, which exchanges with nonradioactive hydrogen in the gas and which would be distributed wherever the gas was used.

Project Ketch was to proceed in five phases. The first consisted of field testing and exploratory drilling. If results of the first phase confirmed that a test could be conducted safely, Phase Two would be undertaken. It included the emplacement, stemming, and firing of the explosive, plus postshot safety surveys. In Phase Three, the experimenters would drill into the chimney, make tests, and attempt to decontaminate the cavity by flushing radioactive gases out of it with nitrogen. Phase Four would consist of building surface facilities for high-pressure testing, and Phase Five would conclude the experiment with plant operations and an evaluation.

Early in 1967, representatives of the AEC and the Columbia Gas System appeared at the state capitol at Harrisburg to sell the experiment to Governor Raymond P. Shafer, the State Advisory Committee on Atomic Energy Development and Radiation Control, and the State Department of Forests and Waters, from which a lease would have to be obtained for drilling and exploding a nuclear bomb in public lands.

Atomic Rebellion

Citizens of Pennsylvania were unaware that the largest and most highly prized public forest in their Commonwealth had been selected by the AEC and Columbia Gas for the first test of a nuclear gas-storage cavity, until an alert reporter for the *Pittsburgh Press* broke the story in mid-February 1967. The public reaction was one of the elements of the test which the experimenters had ignored. Pennsylvanians had been hearing about atomic explosions at the AEC's test site in far-off Nevada or in remote New Mexico, but none had imagined that anyone would attempt to explode a nuclear bomb in the densely populated Northeast, even in a sparsely populated region of it. News of the project was greeted mostly with indifference—except in central Pennsylvania, where a few people began to wonder about its effect on the environment.

One of them was Fred Iobst, a lumberman from Renovo, who had been a state forester in the Sproul Forest. The more Iobst thought about what Project Ketch might do to the forest, the more concerned he became. Drilling preparation for the shot alone would despoil large areas of the forest, he told neighbors, friends, and business associates. Iobst became a powerful foe of the experiment. Everyone who heard him talk about it was impressed by his belief that planting a bomb under a great forest was wrong. It was wrong for the environment, for the hunters and campers who used the forest, and for the people of Clinton County, who lived around it. Iobst attended meeting after meeting to voice his concerns, and before long the ex-forester had become deeply involved in a crusade to save his forest.

Iobst had gone to work as a junior forester in the Sproul State Forest in 1935 after graduation from the Pennsylvania State School of Forestry. Later in life, he had worked as a forester in Virginia and then had gone into the lumber business. He was a natural ecologist, being able to visualize the chain of effects in a forest of even small alterations in its natural setting.

The changes indicated by Project Ketch would not be small, he warned. There would be right-of-way clearance through the trees for a pipeline and it would be all of 200 feet wide. There would be acid pits—deep holes where acid would be poured to soften the underlying rock for drilling. Some of the acid would migrate into streams and kill the fish. In the huge Leidy gas field nearby, acid had leaked from a deep hole and had killed 80,000 fish in Kettle Creek.

"They wanted to blast two hundred and ten cavities under the forest," he said. "And these were taxpayers' lands—they belonged to the people."

Iobst was concerned because "nobody was organized well enough to fight this thing." He put on a sustained and determined campaign of his own, and for twenty-two weeks, he related, he spoke at meetings five nights a week. His telephone bill during that period jumped $62 in the first month. But if any woodsman could have been credited with sparing the trees, it was ex-forester Iobst.

One of the residents whom Iobst influenced was Mrs. Walter Barger of Spring Mills. She conducted an adult discussion group Sunday evenings at the local Saint John's Union Church. It was a small, country church, but well attended, and her group took up a study of "Project Ketch" after hearing ex-forester Iobst. In a few months after the first disclosure of Project Ketch, nearly everyone in Clinton and neighboring Centre countries knew about it and many persons had an opinion, pro or con. But everyone was concerned.

During the summer of 1967, the Governor's Advisory Committee raised the question of radiation contamination of the gas. The experimenters had touched on that problem and then virtually dismissed it in a handsome booklet they published, describing Project Ketch and the benefits it would confer on mankind and Columbia Gas. It was true, the experimenters said, that krypton-85 and tritium might have to be flushed from the chimney, and that gas in the chimney would pick up small amounts of radioactivity. But this contamination could be diluted in the pipeline with gas from other sources, mixed with air for combustion, and the combustion products would be further diluted with ambient air, according to the booklet entitled *Project Ketch*.

"The resulting levels of radioactivity in a room where such natural gas would be burned is predicted to be about at the RPG [Radiation Protection Guide] permissible for continuous exposure to the general population," the booklet assured. It failed to mention, however, that no radiation-protection standards ever had been developed for radioactivity in natural gas. The Governor's Advisory Committee took note of this oversight:

> The entire project is predicated on the feasibility of meeting acceptable levels of radioactivity in gas for commercial use, in the home as well as in industry. We find, however, no reference to work to establish the needed acceptance standards and hence no estimate of the probability of meeting such standards. The maximum permissible concentration values . . . were not developed for this kind of operation and are not believed to be acceptable by the subcommittee.

Nevertheless, Governor Shafer took the position that the Commonwealth would concur in a favorable decision by the AEC that the test could be conducted without harm. The green light was lit for the Phase One feasibility study.

But a small number of voters in central Pennsylvania did not share the official view. Their number was growing. In the fall of 1967, the Clinton and Lycoming County United Labor Councils announced their opposition to underground explosions near Renovo. The Pennsylvania Council of Republican Women issued its Legislative Letter asking: "In considering the Project Ketch proposal, we might ask: what public good can come of this? The private gain is obvious."

The experimenters hastened to explain the public good. Columbia Gas invited twenty-five Centre County opinion-makers, civic leaders, and officials on a junket to Farmington, New Mexico, to observe the Gasbuggy shot on December 10, 1967. The group included county commissioners, state and borough officials, and state legislative assistants. It also included ex-forester Iobst, who returned with the announcement he was more against Ketch than ever.

In the borough of State College in Centre County, site of the Pennsylvania State University, a group of residents, including graduate students and faculty members, organized the "People Against Ketch." The group advertised in the *Centre Daily Times* and invited authors of anti-Ketch letters-to-the-editor, which the newspaper had published, to attend protest meetings.

Leaders of the People Against Ketch were concerned about Ketch, but not necessarily opposed to it until two public forums were held early in 1968 to explain it. The first, which seemed to be managed by the experimenters, was held at Lock Haven State College in Clinton County, where members of the audience were not permitted to question the AEC experts. The second was held on the Pennsylvania State Campus where audience participation was allowed and the discussion lasted far into the night. One of the organizers of the People Against Ketch said that most of the group was neutral on the project until they heard the AEC attempt to explain how safe it was going to be. There was too much hard sell, some of the leaders said, and this served only to increase the concern of the "People" to a state of genuine alarm.

At the Lock Haven State College forum in February 1968, a group of University of Pittsburgh scientists appeared to debate the safety issue. Among them was Dr. Sternglass, then deeply involved in his studies in low-level radiation and infant and fetal mortality. He charged that in at least two-fifths of the AEC's underground tests between 1961 and 1963, radiation had been vented into the atmosphere. Dr. Allen Brodsky of the University of Pittsburgh School of Medicine questioned whether gas stored in a chamber created by a nuclear detonation would be free enough of radiation to be safe for home use.

The Pittsburgh Chapter of the Federation of American Scientists asserted that the experiment involved "serious potential health hazards to the population which have not been adequately explored and evaluated." It said that radioactivity could be released after detonation by ground rupture, by flushing out the cavity around the test site (in this case, Clinton County), by leakage of gas stored under pressure, and by the final delivery of the gas to consumers, adding:

> Accidental release of even a very small fraction of the iodine-131 and strontium-90 generated in the detonation would do irreparable harm, particularly to children who are more sensitive to radiation-induced leukemia and cancer than adults and even to future children of exposed women and girls. Liberation of krypton-85 and tritium could appreciably elevate the long-term contamination of the atmosphere and hydrosphere.

Ex-forester Iobst predicted that preparations for the experiment would mar the forest with access roads, pipelines, sludge, and acid pits. They would endanger deer and other animals, he said, cause the felling of 50,000 trees for right-of-way, open up the forest to soil erosion, disturb underground water channels, and ruin mineral deposits. Besides, the cavity would provide storage for only a fraction of the capacity of the nearby, worked-out Leidy gas field, which could hold 110 billion cubic feet of gas. So who needed Project Ketch?

This meeting marked the beginning of the end for the experimenters. The People Against Ketch were rapidly becoming fully polarized on the issue. They collected more than 9500 signatures to anti-Ketch petitions, dispatched delegations to protest against the project in Harrisburg and Washington, organized a caravan to tour the Ketch site, and sent scores of "We don't want Ketch" letters to newspapers, state legislators, and central Pennsylvania representatives in Congress.

After listening to Iobst describe the damage that Ketch could do to the Sproul Forest, the Renovo Business and Professional Women's club made known its active opposition to Ketch. Representative John P. Saylor, a Republican of Johnstown, joined the anti-Ketch crusade, the first Republican official to do so, asserting:

> The sure way to halt the proposed underground explosion in central Pennsylvania is for the state to insist that the Columbia Gas System Service Corporation, the sole beneficiary of the project, assume full

liability for whatever damages might occur to water, land, property, and people. If the company which is promoting the expenditure of enormous federal funds so that it will have a subsidized reservoir to store its product is willing to put its assets on the line to protect the innocent public, then, perhaps, citizens of the Commonwealth will have a measure of safety assurance that is otherwise absent.

In March, the Centre County Democratic Committee announced it was reaffirming its stand against the project, thus injecting a local, political challenge into what was fated to become much more than a technical issue. Mrs. Marie Garner, the county chairman, explained that the committee felt obliged to speak out because local Republicans were passive on the project.

But the Republicans were simply waiting to see what would happen. Slowly, they began to react to the Democratic prodding. State Representative Eugene M. Fulmer of State College denied the Democratic allegation. Had he not asked a physicist to go in his place aboard Columbia's Gasbuggy junket and report back to him, so that he might have an expert opinion? State Senator Daniel A. Bailey of Philadelphia announced that in spite of what the Centre County Democrats said he was "strictly opposed to Ketch—it would devastate our recreation area." State Representative Max Bossert of Beech Creek denied that he had failed to question the project. He certainly had—but so far nobody had said anything about it to the legislature.

Quis Custodiet

Why were the People Against Ketch against it? One reason was the challenge of the project to their right of self-determination. Some of the People Against Ketch saw in the imposition of a new technology upon their community without their consent a threat to their sovereignty. What right did the AEC, the Governor of Pennsylvania, or a state agency have to despoil a forest belonging to the Commonwealth? Did not such an act require the consent of the people?

This view was expressed by a Pennsylvania State University

botanist, Franklin S. Adams, who raised two questions in a letter to the *Centre Daily Times*. Was Ketch an exercise in decision-making via democratic consent, he asked, or was it "a political boondoggle via an illegitimate mating of the government's scientific enterprise and private industry?" The most serious question raised by Ketch, said Adams, was the fundamental one of who is entitled to make the final decision. He called for a referendum on the issue.

Ketch and its "logical offspring carry us miles and years closer to the ultimate realization of Huxley's *Brave New World* and Orwell's *1984,*" he said.

The "logical offspring" of Ketch referred to an estimate in the *Project Ketch* booklet of a need for 1115 new storage reservoirs, which could be produced by an equal number of 50-kiloton nuclear detonations. The prospect of 55.7 megatons of nuclear explosives being detonated to create gas storage was horrifying—yet this was the prospect which readers of the booklet were led to expect if Ketch were successful. This vision of nuclear damnation of the environment did more to turn people against Ketch than any of the anti-Ketch arguments.

Professor Adams' call for a referendum on Ketch evoked a reply from the Columbia Gas vice-president for engineering and research, Sy Orlofsky. He wrote to Adams: "I happen not to share your opinion that a public referendum on Ketch is desirable. The reason for this is that I believe in a representative government, which we have, rather than a democratic government in which every citizen votes on every issue."

Orlofsky contended that Ketch was "a very technical subject" and thus was "better evaluated and understood by scientific and engineering experts who can bring informed and trained talent to bear in analyzing the matter." An "uninformed vote" on such an issue, Orlofsky claimed, "can very well lead to manipulation and ultimate tyranny."

Adams responded: "Decisions concerning so-called progress that can only be made by so-called technical and industrial experts to the exclusion of the public's right to participate might be better left unmade."

The exchange between Adams and Orlofsky expressed an essential difference in attitudes between the conservationists-environmentalists and the Atomic Industrial Establishment. It was to appear again and again in the citizen protest against both Plowshare and power-reactor technology. The People Against Ketch resented the Establishment telling them what was good for them. The AEC and gasmen viewed the "people" as technological illiterates reacting against "progress" on a primitive level. There was no middle ground. Aaron Druckman, a Penn State professor of philosophy, verbalized another aspect of the "people's" case. "I'm against the use of nuclear explosions in any form and against the use of this technology simply because we have it." he said.

The Fall of Ketch

During the spring of 1968, popular opposition against Ketch grew in the central counties and spread across the Commonwealth. The issues had become clear. In addition to the question of self-determination in the acceptance of a new technology, the project raised questions about the use of public lands for private purposes, the public safety, and the responsiveness of public officials to the electorate. The "People Against Ketch" represented the most vigorous popular uprising west of the mountains since the Whisky Rebellion of 1794 against Alexander Hamilton's excise tax. The central issues of the ancestral and modern protest were not dissimilar.

The anti-Ketch movement expanded through the political power structure of central Pennsylvania rural communities and small towns. The Centre County Pomona Grange sent a note to the county commissioners saying it was not in favor of Ketch. The commissioners subsequently sent a letter to Governor Shafer asking him "to bring all activity related to Project Ketch to an immediate halt." Clinton County Commissioners went on record against the project, stating that "a most disturbing aspect is the reluctance of the persons involved to clearly define what benefits will be derived by the people of Clinton County from an underground nuclear detonation." Lycoming County Commissioners also announced their opposition to Ketch.

The Central Labor Union of Clinton County passed a resolution protesting Ketch. A few weeks later, the Pennsylvania AFL-CIO adopted a resolution at its state convention opposing the project. The resolution, submitted by the Clinton County organization and Local 787, United Papermakers and Paperworkers of Lock Haven, asked: "Is it modern civilization to cause a rapid destruction of water and forest resources through nuclear explosions?"

Anti-Ketch resolutions were reported by the Centre County Federation of Sportsmen's Clubs, the Southern Clinton Sportsmen's Association, the Renovo Lions Club, and the Pennsylvania Division of American Association of University Women, who asked the Governor to retract permission for the experimenters to proceed with Phase One of the experiment.

There had been a hint that the project might provide employment in Clinton County, but it was dispelled when the Pennsylvania Secretary of Commerce, Clifford Jones, said that as far as he could see, Ketch would provide permanent work for only two to four men.

Demands for a referendum and denunciations of the project from various labor, civic, service, recreation, and women's organizations were published almost daily in the central Pennsylvania press. The People Against Ketch appeared before the Public Works Subcommittee of the United States House of Representatives Appropriations Committee to demand withdrawal of projected AEC funds (about $3 million) for the experiment. "Is there no way to stop it?" demanded the Bellefonte Garden Club. The Pennsylvania Chapter of the National Campers and Hikers Association expressed the hope that there was.

There was a way to halt Ketch in its tracks. It became obvious in June 1968 when state office holders, mostly Republicans, realized that the power structure had turned against Ketch, leaving them out on a limb. Congressional Republican representatives climbed on the bandwagon. Centre County's representative in Congress, Albert W. Johnson of Smethport, sent a letter to Senator Pastore on the Joint Committee asking that Ketch be discontinued. Johnson explained: "Because of the widespread distrust of the possible harmful effects of this blast, the people of this area have become bitterly aroused. . . ." He added, as a practical matter: "The large numbers who have indicated their dissent through letters, petitions, and other messages to their Congressmen and to Governor Shafer indicated that it would not be a prudent action to take at this time. Not enough is known about the probable consequences of such a blast."

The Republican organization, which ruled the State House and held most of the political power in the rural central counties, had finally seen the handwriting on the wall. Democrats had been building support all spring for their legislative candidates by using Ketch as an issue. It was distinctly possible that Ketch could be a powerful lever for the Democrats in Pennsylvania in the elections of 1968 and 1970. For the first time, nuclear technology had been thrust into the political arena, where the popular sentiment could be expressed—at least indirectly. No one had planned it that way, least of all the People Against Ketch, but that is how it evolved. As public concern about and opposition to Ketch escalated, the experimenters found themselves increasingly isolated from public officialdom, which at first had given them a warm welcome.

In a letter to the Governor, State Representative Fulmer, who was chairman of the House Public Health and Welfare Committee, finally urged an immediate halt to Ketch, explaining: "The citizens of my legislative district are alarmed and confused to the point where they no longer feel they are being told the truth."

Early in July, the political component of the Ketch controversy reached the gubernatorial level. Milton J. Shapp of Philadelphia, the Democratic gubernatorial candidate who had been defeated by the GOP's Shafer in 1966, announced that he opposed Ketch. There was no longer any question that the Republicans would be saddled with the unpopular side of the Ketch issue if it persisted into the November 1968 state and national elections.

On July 5, 1968, Columbia Gas suddenly announced its withdrawal. The company notified Governor Shafer that it was dropping its proposal to lease the Sproul State Forest site as a result of "re-evaluating" Ketch. Other sites would be studied in other Appalachian areas, the company stated.

So ended, for the time being, the experiment called Project Ketch. It had begun as a technical experiment but had developed into a sociopolitical one of considerable magnitude. Both sides learned a good deal from it. The Atomic Industrial Establishment discovered what had been missing in its feasibility studies—the human equation. The citizens of central Pennsylvania discovered how they could thwart a feared and unwelcome technology by acting through their own power structure, consisting of labor unions, clubs, churches, the Grange, the universities, and local political organizations.

Ketch seemed to be a turning point in the drive of the Atomic Industrial Establishment to devise an energy policy that suited its own interests and to impose it on the United States. From Ketch onward, citizens were to make it increasingly clear that, in the formulation of such a policy, they had to be consulted. But they would not always win.

Rulison

After the fall of Ketch, the Establishment turned once more to gas stimulation in the West. The second big experiment in the development of this technique by the AEC and industry was Project Rulison. It sought to stimulate gas flow in the Mesaverde Formation underlying the 60,000-acre Rulison Field in south-central Garfield County and northeastern Mesa County, Colorado, about forty miles northeast of Grand Junction.

Partners in the experiment were the Austral Oil Company, Inc., of Houston, the AEC, and the Department of the Interior. Program management was provided by CER Geonuclear Corporation of Las Vegas. The experiment called for a 40-kiloton explosion, 8430 feet below the surface on the north slope of an uplift called Battlement Mesa. Scheduled originally for May 22, 1969, the shot was postponed over the summer because of the persistence of heavy snows in the highlands and the likelihood of snowslides triggered by earth shaking. It was reset for September 4, 1969.

Citizen opposition to Rulison took a different course than that in Pennsylvania to Ketch. Rulison was being conducted on private property. State-owned resources were not involved. It was distant from a populated area and, consequently, spontaneous citizen concern was widely diffused. Politically, Rulison was not vulnerable since no state permit was required for the experiment. Environmentalist opposition to Rulison sought relief in the United States District Court at Denver.

Hoping to engage some public support for the test, or, at least, counter opposition, the AEC and its industrial partners brought speakers to explain the project to service clubs and public meetings in the Rulison area. As these preparations were going on, three suits seeking an injunction to halt the testing and, failing that, to prevent the flaring of the gas, were filed in the United States District Court in Denver. One action was brought by owners of property near the drilling site. A second suit was brought by the Colorado Open Spaces Coordinating Council, Inc., on behalf of all citizens, and the third was filed by Martin G. Dumont, the State District Attorney of the Ninth Colorado Judicial District. The suits alleged that the shot would result in widespread property damage from ground motion, would expose the population to the hazard of radiation either from accidental venting or from subsequent flaring of the gas, and would generally constitute a nuisance.

The District Court denied the pleas for a preliminary injunction and was sustained, on appeal by the plaintiffs, by the Tenth Circuit Court of Appeals. The plaintiffs then asked for an injunction to prevent the flaring of the gas on the contention that it would create a health hazard by releasing radiation to the biosphere.

Realizing that they were not only testing a gas-stimulation method, but also public acceptance of it, the experimenters took elaborate precautions to brief everyone who would listen to them on what they expected to accomplish and how it would benefit mankind. They also prepared elaborate safety precautions. For, although they assured everyone that an earthquake or serious venting of radioactivity was virtually out of the question, they quietly prepared for both eventualities.

As the bomb was emplaced in the shot-hole during the latter part of August 1969, a Rulison Information Office was opened in a trailer in Grand Valley, the community nearest the detonation site. Thousands of brochures, pamphlets, and information handouts describing the project were distributed to residents, business people, and tourists. Preparations for such effects as massive landslides were made, consisting mainly of closing roads and evacuating threatened work areas.

On the day of the shot, roadblocks were set up ninety minutes before the detonation, when overflying observation aircraft pronounced all roads clear, until after the shot. The State Highway Patrol, sheriff's deputies, and a sheriff's posse stood by to maintain the roadblocks and lend aid in case of an emergency. No one was really sure what would happen.

The United States Public Health Service sent nurses to visit families living near Ground Zero to arrange for the residents to evacuate their homes on shot day or to have a Health Service nurse stay in the house with the family until after the shot. Gas and electricity were turned off before the shot, as a precaution against earthquake damage. Before the gas was turned on again, it was to be tested for radioactivity. In the town of Grand Valley, residents were asked to stand outside their homes at shot time, to avoid injury in case their houses collapsed.

Damage was expected, but the experimenters hoped that it would be confined mostly to windows, glass jars on shelves, and crockery. The Austral Oil Company assumed responsibility for the first $10,000 in damage claims. Additional settlements would be paid by the AEC up to a limit of $5000 each. Claims in excess of $5000 would be referred to Congress for payment.

Two days before the scheduled shot on September 4, more than six hundred news media and technical observers had assembled in the reception area. The experimenters had a requirement that the winds be out of the northwest, a precaution against the blowing of radionuclides, in case the shot vented, into populated areas. This precaution delayed the shot until the winds were right on the morning of September 10, when the countdown, which was broadcast over the radio, proceeded to zero. About a hundred observers who had weathered the six days of delay peered at Battlement Mesa through field glasses and telescopes as the bomb went off underground. Under their feet, the ground trembled with a shock that registered 5.5 on the Richter scale, a moderate earthquake. Rockslides began on the Mesa, raising clouds of dust. In Grand Valley, windows rattled and houses swayed. There was a fall of loose bricks from the office that housed the AEC damage-claims bureau.

For a time, the telephone system went out, a hiatus reminiscent of the communications blackout in New Mexico when the first atomic bomb was exploded above ground near Alamogordo twenty-four years earlier. But in a short time, all the utilities were turned back on and the telephones were working again. Colorado was still there—as New Mexico had been at the beginning of bomb testing. Roadblocks were lifted. Squads of state police, sheriff's deputies, and possemen roared away in their cars. Traffic resumed and life was normal again.

Deep underground, the cavity formed by the explosion collapsed from the top down after forty seconds to form a chimney estimated to be 301 to 451 feet high, with a radius of 72 to 108 feet. Its volume was calculated at 1.5 to 5.3 million cubic feet.

Radiation from Rulison

In their pleas for an injunction to prevent postshot flaring the gas—burning it at the top of the well to relieve pressure underground—the plaintiffs contended that tritium and krypton-85, brought up from the shot-hole by the gas, were a threat to human, animal, and plant life in the region. Even if the quantities of the radionuclides released were too small to present immediate danger to health, their addition to the cumulative radioactive pollution of the environment would contribute to a future detrimental effect.

Chief Judge Alfred A. Arraj of the District Court denied the plea. In the memorandum of his opinion and order, dated March 16, 1970, he pointed out that the legal question presented by the plaintiffs' allegations was whether the AEC had complied with the Atomic Energy Act and had made reasonable provisions to protect health and minimize danger to life and property. To resolve this issue, he said, two things had to be determined: how much radioactivity would be released by flaring the gas and its effect on health.

The evidence indicated that 960 curies of krypton-85 and from 1000 to 10,000 curies of tritium were expected to be produced by the Rulison bomb. The range in the estimates of tritium depended on the efficiency of a blanket of boron carbide around the bomb in reducing tritium production by absorbing neutrons. If the boron blanket was effective, it would reasonably be expected that tritium production would be about 1000 curies. A gas sample taken just before trial indicated that the tritium produced was about 1300 curies.

Judge Arraj held that probably no more than 2000 curies of tritium were produced by the shot and that the flaring operation would release only 20 per cent of it, or 400 curies. In commenting on the health effect of such a release, Judge Arraj cited the testimony of Dr. Victor Bond, associate director of the Brookhaven National Laboratory, who said that the emission of 2000 curies of tritium a year, as contemplated by the flaring plans, would inflict a maximum radiation exposure of only 2.5 millirems on any person. This was about one-twentieth of the average chest X-ray dose— the same as the radiation dose a passenger would get from cosmic rays during an airline flight from New York to Denver at established jet altitudes. The judge ruled:

> We therefore find that the preponderance of the evidence shows that the Rulison plans for the release from the cavity of gas containing a

maximum of 3000 curies of radionuclides make reasonable provision for the protection of the health and safety of human, plant, and animal life.

Subsequent AEC studies showed that radioactivity in the gas was lower than even the agency had expected. It ranged from the natural background level in the area to about one four-hundredth of the agency's guidelines. No radioactivity above background radiation was detected in streams or in drinking water.[24] There evidently had been no leakage of radioactive material into the underground-water system of the region. Production testing of the new gas well began on October 26, 1970. The well was flared for nine days at rates of from 11 to 17 million cubic feet a day. It was then shut down twenty-seven days to allow pressure to rebuild and flared again, this time for twenty days at a rate of 5 million cubic feet a day. By the end of 1970, the AEC reported, the quality of the gas was improving as fresh, uncontaminated gas flowed into cavity from the surrounding fractured rock.

Public protest in the Rulison experiment had not been entirely in vain. There had been improvement in the "cleanliness" of explosives between Gasbuggy and Rulison. John S. Kelly, director of the AEC's Division of Peaceful Nuclear Explosives, reported that only 1 gram of tritium was produced by the Rulison shot, compared with 4 grams in the Gasbuggy detonation.[25] Even though production testing had not been completed, Kelly told the Joint Committee at its 1972-fiscal-year budget hearings, it was apparent that the Rulison experiment had stimulated gas flow by a factor of five to ten. The first seventy days of flow tests showed production of 400 million cubic feet of gas—more than the first five years of production from nearby wells which were stimulated by chemical explosives.

Moreover, Kelly reported, both Gasbuggy and Rulison had shown that seismic effects could be kept to "acceptable" levels— although the official did not specify to whom the effects would be acceptable. Presumably, the AEC was the judge in this matter. There had been little ground motion from either of the gas-stimulation experiments, he said, and now, after the second shot, damage claims "were primarily for cracks in the plaster or cinder blocks or for bricks dislodged from a chimney."

The AEC published an analysis of Rulison damage claims in February 1971.[26] It reported 557 "credible" damage complaints as of March 31, 1970, including 300 from towns and 257 from rural areas. The complaints cited instances of damage as follows: 143 to chimneys, 148 to interior plaster, 48 to masonry walls, 66 to foundations, 24 to windows, 15 to fireplaces, 21 to other exterior walls, 9 to roofs, 4 to television sets, 18 to other household items, 23 to cisterns, 4 to wells, 8 resulting from earthslides, 7 to utility lines, and 19 to miscellaneous structures. As of October 31, 1970, damage-claim payments totaled $110,167.09 and another $6,358.95 had been offered in settlement of five outstanding claims, but had not at that date been accepted by the claimants. The AEC reported that it had anticipated claims totaling $123,000 in its preshot calculations for the design yield of the explosive. This sum was therefore "acceptable" from the agency's point of view.

Banebury

In the view of the AEC, Gasbuggy and Rulison had proved the feasibility of gas stimulation by nuclear explosives, but critics of Plowshare were not impressed. The danger of accidental release of radioactivity from underground explosions had by no means been overcome. On April 21 and again on December 18, 1970, two underground nuclear tests of undisclosed yield at the Nevada Test Site vented and emitted radioactive debris which was detected far beyond the test-site area. The December 18 test, called "Banebury," resulted in low-level radiation exposure to on-site workers and to nearby off-site residents, according to the AEC's 1970 annual report to Congress. The Associated Press reported that the radioactive cloud from the Banebury Test spread out over an area of Nevada and Utah, between Las Vegas and Salt Lake City.[27] Yet, from an economic standpoint, nuclear gas stimulation appeared promising from the Gasbuggy and Rulison results, if somewhat hazardous to public health. The technique could produce gas for about 30 cents per 1000 cubic feet at the wellhead and return a profit after royalties and taxes.[28]

It was no wonder that the Atomic Industrial Establishment devised a convenient rationale for profits and health hazard in the doctrine of "acceptable" risk. It appeared, however, that the risk-benefit equation worked one way: the residents took the risks and the industry took the profits.

As they reviewed their accomplishments in the spring of 1970, there was a gung-ho feeling among the Plowshare promoters in government and industry. A new deal was looming for gas and oil exploitation. A new era was dawning amid lower-than-expected debris and damage claims from Rulison, which had hardly created much more damage than a small earthquake.

The Division of Peaceful Nuclear Explosives asked for a $36.1-million budget for Plowshare in the 1971 fiscal year, more than double the funding in 1969 and 1970. This sum was pared to $22.9 million by the AEC top management. But it was still considerably more than the annual appropriation in the previous two years of $14.4 million in fiscal 1969 and $14.5 million in fiscal 1970.

But then to the astonishment of most members of the Joint Committee, the Office of Management and Budget (OMB), which acts for the President in over-all budget policy, slashed the Plowshare request to $8 million for fiscal 1971. It was the lowest authorization for Plowshare since the program was first funded in 1961. The gas industry was stunned. Who had pulled out the rug? Why was the program being virtually throttled after a successful test performance? Certainly, the natural gas shortage wasn't easing; it was getting worse.

Greek Fire

Plowshare was cut by the Nixon Administration because its lack of technical urgency and its prospect of becoming a political liability as a conservationist issue made it perfectly expendable. It was apparent that the time for Plowshare had not come. People were not ready to accept it, as Ketch and Rulison had demonstrated on the national scene, and as the shelving of the Panatomic Canal project had demonstrated on the international scene. Moreover, further development of explosives was indicated for both cratering and underground-engineering projects of the future.

Part of the $8-million authorization allowed by the OMB was earmarked for development of more efficient explosives which would yield less tritium. Such an explosive was indeed developed —called "Miniata"—and tested July 8, 1971, in an 80-kiloton explosion 1735 feet under the Nevada desert. Miniata would not only reduce the amount of tritium released but also cut the costs of gas or oil-stimulation explosives. In Rulison, according to Kelly, the AEC had invested $1.2 million and industry, $6.5 million.[29] Future shots would be cheaper.

Still hopeful of better days to come, the Division of Peaceful Nuclear Explosives appeared before the Joint Committee during the 1972-fiscal-year budget hearings with a proposal that it could make nuclear stimulation technology fully available to industry in five years at a cost of only $85 million. But the OMB and Congress had stopped listening to tales of the great bonanzas that could be reaped from bombs. Only $5.2 million was allocated to Plowshare in fiscal 1972, just enough to enable the program to survive at a low level.

The gasmen were disappointed. C.W. Leisk, chairman of Austral Oil, warned that the deficit between discovery and consumption of natural gas was widening.[30] From the Lawrence Radiation Laboratory, a familiar voice was raised to counsel the Joint Committee—the voice of Edward Teller, so-called father of the hydrogen bomb, a title he rejects but one which the press has hung around his neck like the albatross in Coleridge's "The Rime of the Ancient Mariner."

In Teller's view, Plowshare was a tool which a fully mature society could use to manipulate its environment advantageously and thus augment its prospects for survival. There was a historic parallel in the rejection of this technology to the similar rejection of Greek fire, a horror weapon of the Eastern Roman Empire nearly two thousand years ago. Greek fire is believed to have been a mixture of sulfur, naphtha, and quicklime, especially effective in naval warfare because it burned so well on water.

It was this invention of the Byzantines, Teller related, which had turned back the first Arab invasion of the Empire. But even though the weapon was effective, it was considered too horrible for civilized use by many influential people and eventually was outlawed. The result? "Constantinople lost its defense; in the end it fell," he said, adding:

> Greek fire also happened to be the first really impressive use of chemical energy in human affairs. . . . I suspect that suppression of Greek fire, the fact that Greek fire was not only outlawed, but kept secret, delayed the industrial development of the world by almost a millennium. If the discovery of the Greek fire would have evoked more interest and less horror, more openness and less secrecy, the Dark Ages may have been avoided.

Progress cannot be stopped, Teller asserted, and Plowshare will proceed.

"Today," he warned the Committee, "those conservationists who have become reactionary, who are opposed to all progress, who seem to believe that everything that is good lies in the past, may bring about another Dark Age. I hope they will not succeed." [31]

Rio Blanco

And indeed, Plowshare was still a long way from being dead. The Division of Peaceful Nuclear Explosives had another gas-stimulation shot on its agenda called "Rio Blanco." The third demonstration of gas stimulation technology by nuclear explosives, the Rio Blanco project would use three, specially designed charges, which would be exploded simultaneously from 5500 to 6900 feet underground with an aggregate yield of 90 kilotons. It was to be the first test of multiple nuclear explosives, with CER Geonuclear as industrial sponsor.

The scene of the experiment was public land in a sparsely populated region of northwestern Colorado, fifty miles north of Grand Junction and thirty-five miles northwest of Rulison. Detonation of the three explosives was expected to produce a rubble chimney and a region of fractured rock 350 feet in diameter and 1400 feet high.

In its Environmental Impact Statement, drafted in January 1972, the AEC forecasted "minor architectural damage" to 170 buildings in the area, estimated at $50,000. Within a radius of seven miles, less than fifty persons would have to leave the area for a short time, the statement said. Twelve mines within fifty-three miles would be asked to suspend operations during the detonation. While a few dozen "minor seismic aftershocks" could occur, the statement said, the ground motion they might produce would pose no hazard. Flaring of the gas would not be started until short-lived radioisotopes had decayed and the amount of radioactivity released during flaring was expected to be less than that released at Rulison. The draft statement predicted that maximum exposure of local residents to radiation released by production testing would not exceed 1 millirem, less than 1 percent of the annual local background radiation.

If the three-shot test were to prove successful in producing gas without causing more environmental damage than predicted, the AEC proposed moving on to a second phase of the Rio Blanco program and shooting four to six more wells. If these also were successful, it would then go into the third phase of the program calling for twenty to sixty more detonations underground.

"Current speculation," said the statement, "is that the total field may be developed with about 140 wells."

In 1969, the Joint Committee had sponsored a bill in the Congress authorizing the AEC to provide nuclear detonation services on a contractual basis to the gas industry. The bill had been shelved after detailed public hearings, partly because of the Plowshare budget cutback and partly because of doubts about the bill's chances in the light of rising concern in the Congress about the environment. How such an authorization would be implemented under the National Environmental Policy Act had not been clear.

Early in February 1972, the old Plowshare bill was revived. Representative Hosmer assured the House that it had been updated. He described it as "an authorization for the sale of energy to be employed, at least in the near term, in underground engineering projects such as stimulating natural-gas flow in deep, low-permeability rock formations." It would conform to the NEPA.

Plowshare was still a going concern, though greatly reduced, in 1972. It still offered the gasmen a chance to shake the gas out of the rocks beneath mountains, deserts, and plains. But the environmental movement was a going concern, too. Another battle loomed in the Old West.

9
The Battle of the Breeder

"Without the use of atomic energy, it is impossible to picture a continuously developing civilization on earth," said Lev A. Artsimovich, chief of thermonuclear research in the Soviet Union. "After all, the stocks of mineral fuel are not boundless. Moreover, there is the increasing problem of air pollution in towns and cities as a result of burning such fuel." [1]

A spokesman for the Atomic Industrial Establishment in the United States could not have stated the case for atomic power more clearly. It is the rationale for the development of atomic-power plants in the entire community of nations using nuclear energy, and it has been particularly applicable to the development of the next generation of fission reactors, the breeder.

Members of the AEC and of the Joint Committee have repeatedly cited the advantages of the breeder: it offers a huge saving in fuel costs to the nuclear-power industry; it promises to extend the reserves of uranium by thousands of years; and more efficient than the present generation of light-water reactors, it will release less waste heat to the environment. Some members of the Joint Committee look at it as a lifesaver technology. Soliciting support for funding the breeder from his colleagues, Representative Hosmer warned in a letter June 29, 1971: "It is an incontroverted fact that our country is running out of power. Something has to be done to avoid brownouts and blackouts. We have a plan. It is to develop a breeder reactor which will safely produce electricity and more nuclear fuel than it consumes."

Whereas present power reactors represent the Model T phase of

The Battle of the Breeder / 233

evolving nuclear technology, breeders represent a Model A phase. Their biggest advantage over the light-water-moderated reactors now in use is their ability to convert nonfissionable uranium-238 into fissionable plutonium-239, which is atomic fuel. Moreover, the amount of plutonium bred from U-238 in a fast reactor averages one and one-half times the amount of fissionable U-235 required to breed it.

Beyond the Model A phase of a nuclear-fission reactor is a more modern machine, the controlled thermonuclear fusion reactor. (The uncontrolled version of fusion is the hydrogen bomb.) Most nuclear engineers and physicists regard fusion as the ultimate source of energy to meet mankind's requirements for the next several million years.

If that is the case, why have the AEC and the Joint Committee pulled out all the stops for the breeder reactor? The answer is that the breeder is here—the technology is sufficiently developed to be ready for a commercial-power demonstration in three to five years and then for the power market a few years later. But the fusion reactor is technically still around the corner. So far, no one has been able to design a fusion reactor that is likely to be economic. It is not for lack of trying. Engineers and scientists in the United States, the United Kingdom, and the USSR have been working on a fusion reactor for more than twenty years. In recent years, the whole nuclear world has been working on the problem, virtually with the coordination of an international team. Since the end of 1970, most researchers have agreed that a workable fusion reactor can become a reality in this century. But when, exactly, depends on the level of research and development funding, and this has been depressed in the United States by the all-out drive of the Atomic Industrial Establishment to bring the breeder on line.

Fusion offers a high probability of converting the energy of nuclear reactions directly into electricity with an efficiency of 90 per cent or better without going through the steam cycle. The flow of charged particles through the reactor, which might be as big as a football stadium, would be harnessed and fed into a wire as electricity. Otherwise, the fusion reactor would be used as a heat source, like a fossil-fuel plant or a fission reactor, to boil water, make steam, and drive a turbine. Heat in the fusion reactor would be produced mainly by collisions of high-energy neutrons with a surrounding material blanket, possibly of lithium. It would be conducted by a coolant to a heat exchanger, as in a fission reactor.

In contrast to fission, which results from the relative instabilities of the heavy elements (uranium, thorium, or plutonium), thermonuclear fusion depends on the light elements for terrestrial applications—particularly the hydrogen isotopes deuterium (H^2) and tritium (H^3). The fusion of two deuterium nuclei (the D-D reaction) produces a helium nucleus and a neutron plus energy of 3.25 million electron volts. Another kind of D-D reaction produces a tritium nucleus (triton) and a hydrogen nucleus (proton) plus 4 million electron volts. The triton then reacts with another deuterium nucleus (deuteron) to form a helium nucleus and a neutron, with an energy release of 17.6 million electron volts.

Deuterium exists in the ratio of 1 atom of H^2 to every 6500 atoms of ordinary hydrogen H^1, and is relatively inexpensive to recover as fuel. Its potential energy is enormous. The amount of deuterium in a gallon of water would yield energy equivalent to that of 300 gallons of gasoline, according to one authoritative estimate.[2] The deuterium in the world's oceans offers an unlimited energy resource, at least enough for many millions of years.

The Reluctant Nuclei

Achieving the fusion bonanza has been a long, hard road of disappointment and frustration. It is beset with the difficulties of perfecting a technology that will force two reluctant nuclei together —either two deuterons or a deuteron and a triton. These nuclei repel one another because each carries a positive electrical charge. Consequently, a considerable amount of energy must be applied to overcome this electrostatic repulsion.

In stars the energy required to fuse the nuclei is supplied by gravity and heat. On earth, energy must be provided to heat the hydrogen isotope fuel to a temperature of 100 million degrees. In the hydrogen bomb the heat is provided by a fission bomb trigger. In controlled nuclear fusion, however, where the fuel must be at extremely low densities to avoid the development of high gas pressure, the heat may be generated by passing an electric current or a laser beam through the gas, which is under compression by a magnetic field.

At very high temperatures—millions of degrees in a fusion chamber—gas becomes ionized. This is a state wherein electrons have been divorced from atomic nuclei and each is going its own way as a free agent, the electrons being negatively charged and the nuclei carrying a positive charge. In this state the gas no longer behaves as gases normally do. It exists in a fourth state of matter —being neither a liquid nor a gas, and certainly not a solid—a state called a plasma. If the bare nuclei in a plasma can be excited to a high enough energy and if they can be confined for a short time, many will fuse when they collide, producing the thermonuclear reaction that releases nuclear energy.

At fusion temperatures, plasma can be contained on earth by only one means known to science—a magnetic field, which influences the behavior of the charged particles. Contact with ordinary matter causes the particles to lose energy, reducing the temperature of the gas by heat conduction.

For the last twenty years, the main effort in fusion research has been to design a magnetic "bottle" capable of containing the plasma long enough and at densities high enough to ignite a fusion reaction and keep it going for a tenth of a second. At that confinement time, plasma at a density of 1000 trillion particles per cubic centimeter (which is $1/100,000$ of air density at sea level) and at a temperature of 100 million degrees will produce more power from the fusion reaction than is needed to initiate it.

During the early 1950s researchers in the United States (Project Sherwood), the United Kingdom, and the USSR worked independently and under veils of secrecy to solve the problems of plasma confinement. But the answers eluded them, and in 1958 the nuclear powers agreed to pool their efforts for the benefit of mankind—a lofty motive which by that time the competing powers felt they could well afford. Since then, encouraging progress in controlling plasma instability has led to increases in both confinement time and plasma densities so that a controlled nuclear fusion device which would yield more output than input energy appeared feasible. The advances in containment were made a series of devices, starting with Russia's Tokamak, the Oak Ridge National Laboratory's Ormak, Princeton University's Stellarator, Los Alamos's Scylla, and the Lawrence Radiation Laboratory's 2-X.

By 1971 the pitch of international cooperation in the problem of plasma confinement had reached the level of a global commitment. New ideas were eagerly, sometimes ostentatiously, paraded at such meetings as the Fourth International Conference on Plasma Physics and Nuclear Fusion Research, held in June at the University of Wisconsin, Madison, and the Fourth International United Nations Conference on the Peaceful Uses of Atomic Energy, held in September in Geneva, Switzerland.

It was a big year for fusion, a turning point. At both meetings, the experts expressed confidence that the problem of confining plasma was being solved from several directions. What remained to be established was the most feasible method—in terms of power cost. Two factors would determine the pace of fusion-power realization: one, the amount of research funding; the other, the intensity of the social demand for completion of the fusion technology. The two conferences demonstrated the most extensive cooperative research effort on an international scale in the history of technology. Summarizing the situation in the spring of 1971, Richard F. Post of the Lawrence Radiation Laboratory told the National Academy of Sciences that while fusion power had not yet arrived, there existed "A well-grounded, world-wide research effort specially aimed at achieving it, plus recent evidence that says the research effort is on the right track." [3]

The partial success in confinement had shown not only the way to go, but what the problems really were. At the Wisconsin Conference, Roy W. Gould, who was to become head of nuclear research for the AEC a few months later, said that there was a consensus among researchers that it would require a doubling of the 1971 research-funding level to develop a workable reactor within this decade. Kenneth Fowler of the Lawrence Radiation Laboratory explained that experiments have become larger and more expensive as research goals have become more imminent and as the development path became more clearly defined. Besides containment, there were other unsolved problems. Chief among them was the amount of damage neutrons would do to the crystal structures of vacuum-chamber walls and to the magnets.

Safer than Fission

In principle, fusion power offers a cheap, virtually unlimited source of energy, with minimal stress on the environment. What are minimized are two sources of pollution from fission reactors —thermal pollution and radiation. A fusion reactor which converts energy directly into electric current virtually eliminates waste heat, as theoretically it has an efficiency of 90 per cent or higher. This is better than twice the efficiency of the most advanced fission reactor, the Liquid Metal Fast Breeder, for which only 40-per-cent efficiency is claimed, meaning that 60 per cent of the heat energy produced by the breeder is wasted—dissipated into the environment as thermal pollution.

Moreover, the fusion reactor contains only a small quantity of fuel at any given moment. There is no possibility of a core meltdown that would scatter radioactive nuclides over many square miles, as there is in all the fission-reactor types. Nor is there any possibility of plutonium fuel reaching critical mass and blowing up, as some critics contend exists in the breeder reactor. Since the fusion reactor does not use uranium or plutonium, there would be no problem of radioactive-waste disposal. The principal radioisotope that can be emitted by the fusion reactor is tritium, which engineers believe can be totally confined. Thus, since this type of power plant cannot blow up, run berserk, melt, spew radioactivity over the countryside, and pollute rivers and streams with waste heat, fusion reactors offer a safer and more reasonable way of using atomic energy—at least they appear to do so at this stage of their development.

When will fusion be here? The enormous investment of government and industry in fission reactors guarantees that fusion power, even if technically around the corner, won't be available for a long time. There are simply no economic or political reasons for accelerating fusion development. There are only assumed environmental advantages and, thus far, these have not carried great weight with the Atomic Industrial Establishment. Because of the promising fuel economies offered by the breeder, there is little profit motivation in pushing fusion . . . and too many uncertainties to lure a great deal of industry's investment capital. Artsimovich summed it up when asked at the Geneva conference when he expected thermonuclear fusion power to become a practical reality. "When the world really needs it," he replied.

Sunshine and Steam

Besides supporting fusion research, environmentalists have urged the AEC to examine other power options: solar power and the exploitation of natural steam and hot water heated by radioactive processes in the earth. The agency, however, is concerned only with atomic energy. While the potential of these energy options has been scanned by the Federal Power Commission, there is no big-scale effort by any agency of the federal government to develop them.

The FPC has estimated that the amount of solar energy striking the atmosphere each year equals that in 200 trillion tons of coal.[4] About 61 per cent of it reaches the ground, where it becomes transformed as the biological energy source for the planet's flora and fauna. Small-scale applications of solar energy have been developed in Israel in the form of fireless cookers and roof hot-water heaters that collect and focus sunshine on a target. This principle is also applied to agricultural irrigation pumps, in which sunshine is focused to heat a working fluid. But the large-scale application of sunshine reaching the ground to the production of electricity has remained in the realm of exotic power ideas. We use gravitational energy in the form of falling water to generate electricity, but the exploitation of solar energy apparently requires a highly sophisticated optics technology that remains undeveloped, although scientists have been working on the problem.

Two investigators, Norman C. Ford and Joseph W. Kane of the University of Massachusetts, have proposed a method of using sunlight to produce hydrogen cheaply enough so that it could be sold as fuel in competition with natural gas.[5] They calculated that sunlight collected over an area of two square miles by arrays of Fresnel lenses (500,000 of them) would heat water in a larger boiler to 1500 degrees centigrade. As at this temperature about 0.07 per cent of the water vapor thermally dissociates into hydrogen and oxygen, the hydrogen could be pumped out of the boiler through a membrane of palladium, which is permeable to hydrogen, and stored, and the oxygen could be collected, tanked, and marketed. Excess steam would be applied to steer the lens array so that it would follow the sun across the sky during the day.

Another proposal by University of Arizona's optics expert, Aden B. Meinel and his wife, Marjorie, is to focus sunlight on films with a selective surface capable of passing the sunshine but trapping infrared radiation. Again, the collecting area would be large, about 3.8 kilometers (2.36 miles) on a side. The collected heat would boil a fluid to produce steam to turn a turbine, with an assumed collection-conversion efficiency of 30 per cent. A series of these collectors, or "energy farms," located in a sunny area of 78 square miles could produce 1 million megawatts of electricity. The Meinels have suggested the border region between Arizona and California or the Rio Grande Valley for large-scale solar-energy "farming."[6]

In the opinion of John S. Nassikas, Federal Power Commission chairman, an efficiency of 30 per cent is about as high as one might expect from solar-energy power.[7] At that efficiency, however, he estimated that each square mile of the earth's surface holds the potential for 1000 megawatts of power—the amount of power produced by the largest light-water reactors, operating at efficiencies of less than 40 per cent.

Other methods of converting sunshine into electric power have been considered by the FPC. Large-scale deployment of solar cells that convert sunshine directly into electricity might be developed if the cost of the cells could be reduced. At present, they are confined to satellites, where cost is not an inhibiting factor. A National Aeronautics and Space Administration official, William R. Cherry, head of the Space Power Technology Branch, has proposed a "solar rug" consisting of a square mile of solar cells. It would be floated in the atmosphere at an altitude of 50,000 to 70,000 feet and would produce 250,000 kilowatts of direct current. The device is conceived of as a manned power station, supported by a helium mattress or balloon and tethered to the ground by a cable that also might conduct the power to the surface for distribution. A variation of this scheme is a solar-cell array in a synchronous earth orbit, where it would appear to hang stationary over one point on the surface, beaming power by microwave to a ground station.

Full Steam Ahead

Underground, especially in the West, lies a vast supply of natural steam, a source of power which is only partially tapped. Robert W. Rex, assistant director of the Institute of Geophysics and Planetary Physics at the University of California at Riverside, has estimated that the potential heat energy in underground hot water and dry steam in the West ranges from 100,000 to 10 million megawatts.[8]

Where ground-water temperatures are high and the pressure is low, the water boils. Dry steam accumulates in fractures and pores of the subsurface rocks. It can be tapped by drilling wells. In geysers, its pressure is often 100 pounds per square inch. The natural steam could be piped into turbines to produce electricity. In northern California, Rex estimated, geyser-steam fields have a potential of 512 megawatts of electrical power.

The steam fields are not considered temporary. Such fields in Italy have been producing steam for thirty years. According to Rex, some fields in the American Southwest may be expected to produce for 3000 years.

In addition to dry steam, there are extensive fields of hot water —some as hot as 700 degrees Fahrenheit—estimated as twenty times more extensive than the dry-steam fields. In northwest Sonora and northern Baja California, Mexican authorities estimate the capacity of hot-water geothermal power at more than 100,000 megawatts. American potential in the Imperial Valley, a major rift valley with a high heat flow, is estimated by Rex at 20,000 to 30,000 megawatts. Moreover, Rex calculated that these fields would yield 5- to 6-million acre feet of water a year, about one-half the flow of the lower Colorado River. This distilled water, he said, could be used handily to reduce the salinity of that river.

Another source of electricity is a variation of the conventional generator in which a metallic conductor rotates in a magnetic field. Instead of the conductor, a hot, high-velocity gas is jetted through the magnetic field producing direct electric current.

Like the conventional generator, a source of energy is needed to heat the gas to temperatures of 4000 to 5000 degrees centigrade in the presence of an ionizing agent such as cesium or potassium. The gas then becomes a plasma, with electrons separated from atomic nuclei. When electrodes are introduced into the plasma, direct current is drawn off. The process is called magnetohydrodynamics, or MHD.

Compared with conventional fossil-fueled, steam-cycle generators, with 40 per cent efficiency at most, the MHD generator offers 50- to 60-per-cent efficiency—provided the excess or exhaust heat is used to run a steam-cycle generator or gas turbine in addition. Otherwise, a great deal of waste heat is lost to the environment. Unless this process is used in a fusion plant, MHD does not resolve the problem of air pollution.

Back to the Breeder

In the context of the present American power economy, these "exotic" sources of power are not fated for any extensive development as long as the national power policy is dominated by the owners of fossil fuels conventional power-plant interests, and the Atomic Industrial Establishment. I have digressed to discuss fusion and exotic power possibilities to point out that options do exist other than the development of a more advanced and possibly dangerous fission reactor.

So far as national policy is concerned, the breeder reactor represents the shape of the advanced power technology which has been adopted in the United States, Great Britain, Western Europe, and the Soviet Union.

The era of the breeder in the United States was formally proclaimed June 4, 1971, by President Nixon in a special message to Congress:

> Our best hope today for meeting the nation's growing demand for economical, clean energy lies with the fast-breeder reactor. Because of its highly efficient use of nuclear fuel, the breeder reactor could extend the life of our natural uranium supply from decades to centuries, with far less impact on the environment than the power plants which are operating today.

The President called for the construction of a Liquid Metal Fast Breeder demonstration plant, a reactor type which is favored over other types by the AEC, to be sited either in New York or eastern Pennsylvania. Later, on July 26, the President in an appearance at Hanford, Washington, said he would seek Congressional authorization for a second demonstration plant—which probably would be located there.

President Nixon's endorsement of the breeder gave it the highest priority in the AEC's development program. The construction of two demonstrators, rather than the one originally planned, was demanded by the supplier industry as a means of guaranteeing fair competition. As noted earlier, three suppliers were in the running to build the first Liquid Metal Fast Breeder Reactor (LMFBR)—North American Rockwell, General Electric, and Westinghouse. A clamor for a second demonstrator, also funded principally by the AEC, was raised at the American Power Conference in Chicago in April 1971. In the long run, argued John W. Simpson, president of the Westinghouse Power Systems Company, it would be cheaper to build two demonstrators than one. For true competition, he said, two companies building light-water reactors (General Electric and Westinghouse) were better than one. Power men were generally in sympathy with this view. A single supplier given the contract to build a lone demonstrator would certainly get a leg up on breeder technology, with the aid of the government. Two demonstrators, each being built by a different supplier, would serve to keep the industry honest, not only on cost but in sharing technical information. The argument was certainly persuasive and President Nixon bought it.

In espousing the cause of the breeder, however, the President acquired a highly articulate group of critics as his probable foes in the election of 1972. It was the hope of some of them that the breeder program could be made an issue in the campaign on the grounds that the breeder would generate a plutonium-fuel economy and the widespread use of plutonium would be a potential menace to the health and safety of the whole population.

Clementine

The LMFBR which was sold to the President as the way out of a future power dilemma was evolved in parallel with its less efficient predecessor in commercial application, the light-water reactor, LWR. About $650 million has been invested in breeder development by the AEC over a span of a quarter of a century. An experimental reactor called "Clementine" was used at Los Alamos from 1946 to 1953 in early investigations of the problems of operating a machine with fast neutrons (rather than neutrons slowed down by water in the conventional light-water reactor) and a coolant of liquid metal (sodium) instead of water. However, the first, full-fledged Experimental Breeder Reactor (EBR-I) was built at the National Reactor Testing Station in Idaho. In December 1951, it produced the first electricity from a nuclear reactor—enough to turn on electric light bulbs in Arco, Idaho, for about an hour. A larger experimental reactor, EBR-II, which went into service at the Idaho testing station in 1963, has produced up to 20,000 kilowatts of electricity, according to the AEC. EBR-II has served as a test bed for data on sodium as a coolant.

In addition, a small, sodium-cooled demonstration breeder was built at Lagoona Beach, Michigan, by the AEC and the Detroit Edison Company in 1963, amid criticism by some AEC personnel that the technology was not sufficiently advanced to justify such a demonstrator—especially in a populated area. This reactor, called Enrico Fermi I, was designed to produce 67 megawatts of electrical energy, but did not reach the design output for seven years, partly as a result of a partial meltdown in 1966 caused by a mechanical blockage of the liquid-sodium coolant—an incident mentioned earlier, in chapters 1 and 6.

Both EBR II and Fermi I proved that the LMFBR poses some stiff technical problems and will require a much longer shakedown period of testing and adjustment than the light-water reactors needed. Nevertheless, its potential advantages seem so attractive to both industry and government that the LMFBR represents to them the salvation of a civilization which appears to require ever-increasing amounts of electric power. As early as 1962, the AEC predicted, in a report to President Kennedy, that the breeder was the answer to the problem of cheap energy for the future. At that time, it was generally believed that fusion power was too far in the future even to consider.

Removing the Inhibition

The cost advantage of the LMFBR, or any type of breeder, is a result of its more efficient use of neutrons, which not only produce fission in fissionable isotopes but also convert those which are not fissionable to those which are. So far, the only fissionable material which man has found in nature is uranium-235, an isotope which is not abundant and which is very expensive to separate from the much more abundant but nonfissionable uranium-238.

As I have mentioned earlier, the boiling- and pressurized-water reactors now in commercial use in the United States burn U-235. A few of the neutrons produced by the fission reaction, in excess of those needed to maintain it, bombard U-238, which is part of the fuel mixture, and convert it to plutonium-239—a fissionable isotope. Most of the excess neutrons in light-water reactors, however, are slowed down or "moderated" by the water. This avoids overheating and also inhibits U-238–Pu-239 transmutation.

If the moderator is removed and the neutrons allowed to retain their high speed, a great deal more U-238 will be bombarded and transmuted into plutonium than if water or graphite are present.

In the fast breeder, there is no moderator to inhibit the neutrons. That is why it is called "fast." Consequently, a large amount of plutonium is "bred," which is why it is called a breeder. Something else happens. Since two or three neutrons are emitted in the fission of each uranium nucleus and only one is needed to carry on the reaction, the chances are that more than one neutron will impact the U-238—thus creating more than one unit of Pu-239 from each unit of the fissioning fuel. On the average, 1.2 grams of new plutonium are created from every gram of fissionable fuel. That is the basis for the proud claim that the breeder produces more fuel than it consumes. It certainly does, but that does not mean it gives something for nothing. What is used up is the U-238. However, there is so much more of it than the U-235 on which the light-water reactors now rely that by breeding and burning plutonium, the breeder reactor can extend uranium resources from hundreds of years to many thousands of years. Ultimately, of course, all of the U-238 that can be mined would be used up in the conversion to Pu-239, and eventually all the plutonium would be burned. So would thorium-232 which, like U-238, is not fissionable but which can be converted by neutrons to U-233, which is fissionable. But the time when mankind would be expected to run out of uranium and thorium is too distant to worry anybody today.

Breeding Dangerously

The present concern is that the LMFBR technology exacts a price for the increase in neutron efficiency and fuel economy. Some conspicuous difficulties were noted in the November 19, 1971, issue of *Science* by Allen L. Hammond.[9] One is the problem of handling highly reactive sodium, which ignites in the presence of air or water, and in manipulating the extremely toxic plutonium fuel. Sodium in the primary coolant loop passes through the reactor core and becomes radioactive. The heat it picks up must be transferred to an intermediate heat exchanger containing nonradioactive "clean" sodium which, in turn, flows to the steam generator. The sodium must be kept exceedingly pure, according to Hammond, because even small specks of impurities can obstruct its flow or cause it to become corrosive so that it eats through the plumbing.

In the fuel core, the fissionable material will probably be a mix of uranium oxide and plutonium oxide, formed into small pellets which would be packed in thin rods of stainless steel. Additional rods containing U-238 for breeding would be located in the core blanket. Hammond noted that the fast neutrons emitted in the core can cause radiation damage in the steel jackets of the fuel, creating small voids. The voids grow and cause the steel to swell, become brittle, and break. This effect, discovered only a few years ago, would decrease the economic expectations of the reactor by increasing the doubling time of the fuel—the time required for the reactor to double the amount of fissionable material originally present. A doubling time of six to twenty years has been predicted. The industry hopes to achieve a doubling time of less than ten years.

Beyond these problems looms the more menacing one of managing the plutonium—preventing it from polluting the environment when fuel rods are taken out of the reactor for reprocessing and keeping track of the plutonium to avoid loss or theft.

The Plutonium Panacea

The dazzling prospect of a plutonium-power economy was outlined in glowing terms in the autumn of 1970 by Seaborg, in a speech at Santa Fe, New Mexico.[10] He had a right to be enthusiastic about it. After all, Seaborg was a co-discoverer of plutonium and had shared the 1951 Nobel Prize in Chemistry for that feat with Edwin M. McMillan. Looking ahead, Seaborg predicted that by the mid-1990s, the LMFBR would become the predominant type of nuclear-power reactor, supplying two-thirds of all nuclear power in the nation. By the year 2000, he went on, nuclear-power plants will be generating 70 per cent of the nation's electrical energy requirements, with breeders supplying the vast bulk of that portion. Seaborg envisioned a plutonium-powered artificial heart, as well as plutonium-powered spaceships exploring the outer planets of the solar system.

Plutonium would be the great energy base for the American future. As Seaborg saw it, the Gross National Product would rise from $932 billion in 1969 to $3 trillion in thirty years. Basic to this growth was the increase in electrical energy, he said, and basic to the increase in electricity was plutonium, "a vital element of the over-all economic well-being of this country."

To Seaborg and his colleagues on the AEC and to some members of the Joint Committee, the plutonium-fuel economy would fulfill a dream of opulence. But to a number of scientist critics of the AEC, it was a nightmare. Pointing out that plutonium is toxic beyond human experience, Donald Geesaman, Tamplin's associate at the Lawrence Radiation Laboratory at Livermore, warned it was "demonstrably carcinogenic to animals in microgram quantities," stating that injection of a millionth of a gram has caused cancer in mice and dogs.[11] Moreover, he said, "our transition to plutonium as a major energy source will inextricably involve our society with the large-scale commercial production of a substance that is a suitable nuclear explosive." Geesaman estimated that by the year 2000 annual production of plutonium would exceed 100 tons. Can such quantities be protected from internal subversion? Underworld involvement in the trucking industry is legendary and theft epidemic, he said, adding that university unrest is ubiquitous and radical activism a reality.

AEC Dissent

Long before President Nixon proclaimed the Liquid Metal Fast Breeder Reactor a priority, the AEC's chief of reactor development, Milton Shaw, had established it as a top-priority AEC goal. Scientists working on other breeder-reactor types, such as the gas-cooled and molten-salt fast reactors, felt that Shaw's emphasis on the sodium-cooled machine tended to focus the attention of Congress and the White House on only one type, even though the other types seemed to them to show equal promise.

The main effort for the AEC's LMFBR development operations was the work done at the Argonne National Laboratory, which has an experimental site at the National Reactor Testing Station in Idaho. Shaw re-emphasized the importance of the LMFBR in the Argonne program, but changed its control from the laboratory's autonomous and comprehensive one that had hitherto paced LMFBR development throughout the country, to an externally directed program closely managed by Shaw himself. Among some Argonne experts there was dissent to Shaw's detailed direction, but there was no opportunity in the context of laboratory policy to express it. In spite of the fact that the laboratory was operated by the University of Chicago and a cluster of other institutions forming the Argonne Universities Association, the reactor-research program had become geared to LMFBR development priority. This industrial style of research and development represented a 180-degree turn away from more imaginative, experimental work for which the laboratory had become highly respected. In the opinion of one astute observer, there was a consensus among staff members that the narrowing focus of research activity was "debilitating to the institution and destructive to the best interest of science." Even those accustomed to applied work resented the "burden of closely directed research," according to the observer's report.[12]

How this emphasis on breeder-reactor development served the interests of the universities contractually managing Argonne for the AEC was mysterious. But the program under Shaw's direction did serve the interests of the Atomic Industrial Establishment, which in turn insisted that it served the best interests of civilization.

A powerful economic rationale for breeder development could be cited. It was Seaborg's contention that at $10 a gram, the annual value of plutonium recovered from the light-water reactors alone would increase from $4 million in 1970 to $150 million by 1980.[13] By 1990, the total value of all the plutonium fuel in existence would be about $6 billion, or twice as much as the $3 billion which the investor-owned electric utilities spent for all the fuel they burned in 1969.

Another aspect of the economic urgency to develop the breeder promptly was the prospect of a rapidly escalating price of uranium-oxide concentrate. As the more easily accessible uranium was consumed, the more costly it became to mine. At the expected rate of increase in nuclear power, the price of uranium oxide would be expected to jump from $8 a pound in 1971 to $15 a pound by the year 2000, according to an estimate presented to the National Academy of Sciences by Manson Benedict of the Massachusetts Institute of Technology.[14] Since the cost of electricity from light-water reactors increases by six-hundredths of a mill per kilowatt hour with each dollar a pound increase in the price of uranium, Benedict said, the rise in fuel cost would add more than four-tenths of a mill per kilowatt hour to the cost of power. "Even more serious, of course, would be the complete exhaustion of all of our presently estimated resources of low-cost uranium in less than thirty years," he added.

The M.I.T. expert said that because the fast breeder uses relatively little natural uranium, it could be fueled economically with uranium costing as much as $50 to $100 a pound. He cited an AEC report showing that at $100 a pound, United States resources of uranium are estimated at 25 million tons. At $15 a pound, the resources are estimated at 1.5 million tons, and at $8, only 594,000 tons.[15] However, fueled with uranium costing $100 a pound, Benedict projected, fast-breeder reactors could provide all the electricity which is presently generated in the United States for 64,000 years. "It is this tremendous extension of our fuel resources which makes the development of the breeder reactor so challenging and important," he said.

Intervention

Another group of scientists, however, disagreed that the fast breeder was either urgent or important to future American civilization. On May 25, 1971, ten days before President Nixon delivered his energy message proclaiming the LMFBR a national priority, a suit was filed in the United States District Court of the District of Columbia asserting that the AEC was proceeding illegally with its $2.5-billion breeder-reactor program. The suit was dismissed a year later, but the issues it raised persisted, expressing the basic, environmentalist protest against the breeder.

The plaintiff was the Scientists' Institute for Public Information (SIPI), an organization of thirty-five scientists that had pioneered in warning the public of the dangers of radioactive fallout from nuclear-bomb testing in the atmosphere. Margaret Mead, anthropologist of the American Museum of Natural History, was its president, and Barry Commoner, plant pathologist and pioneer environmentalist of Washington University, St. Louis, was chairman of the Board of Directors. SIPI was represented in the action by staff attorneys of the Natural Resources Defense Council, an alliance of environmental-protection organizations, with headquarters in Washington, D.C.

The complaint charged that the AEC was violating the National Environmental Protection Act in that the agency had failed to produce an environmental-impact statement on the breeder program and to list alternatives.

The suit was the second challenge to the AEC on the broad issue of the agency's obligation to implement the NEPA. At the time it was filed, Maryland citizens, the National Wildlife Federation, and the Sierra Club were awaiting judgment from the United States Circuit Court of Appeals in the District on the issue of whether the AEC's rules affecting nuclear-reactor licensing adequately implemented the NEPA. This case, destined to produce a landmark decision, will be discussed in the next chapter. In both actions, the common denominator was the effort by critics of the AEC to breathe some life into the NEPA, which otherwise might lie in suspended animation for generations, like the Refuse Act of 1899, as a result of bureaucratic indifference and the lack of a will to enforce it. Where was the Environmental Protection Agency in this issue? It remained passive.

In a background statement explaining its intervention in the AEC's breeder program, SIPI stated it had simply asked the court to require the AEC to consider alternatives to its program of development and proliferation of a new type of nuclear-power plant. "The important first step is for the AEC to fulfill the legal requirement of the NEPA by issuing a statement describing these alternatives and the potential impact of this program on the environment," the statement said.

The suit essentially challenged the introduction of a new technology by a governmental agency without an accounting of its environmental effects. Although light-water reactors have drawbacks, SIPI stated, their design has acquired important safety features. Water moderates the neutrons which split the atoms of U-235, and any overheating or mechanical disturbance would expel the water and cause the reactor to shut down. But the LMFBR, the statement continued, does not slow the neutrons. Overheating or an outside disturbance might result in compacting the fuel, causing an acceleration of nuclear reactions which could produce an explosion, releasing enormous quantities of radioactive materials, including plutonium, to the environment. Even in the event that such an accident never occurs, the statement went on, "a large number of LMFBRs may result in plutonium contamination of the environment." It cited Seaborg's prediction that in thirty years LMFBRs would be generating a large share of the nation's electric power.

Aside from the safety issue, SIPI challenged the AEC dogma that the LMFBR was an essential power adjunct to the future of civilization. The scientists' statement argued that: "It is by no means necessary or certain that electric-power production will continue to grow at present rates. The limited ability of the environment to absorb waste heat and pollutants will compel a slowing or even a reversal of this growth."

If the desirability of rapid and continued power growth is questionable, SIPI stated, the development of the LMFBR, which has its main justification in continuing this growth, would "therefore seem to be open to serious question."

SIPI thus suggested that instead of meeting an inevitable need, the AEC's breeder program was creating a supply which would stimulate a need, a reverse order of things not unknown in American power economics. For generations, the electric-power companies had been stimulating a need for greater power consumption by promoting the marketing of devices which use power—electric space heating is a notable example—and by establishing promotional rates which reduce the cost of units of power with increased use. From this point of view, increasing power consumption was being induced by the electrical industry as a means of insuring its own expansion, irrespective of social needs and environmental effects. These were implications of the AEC power program which the SIPI statement raised as national issues. By keeping its mouth shut on these matters, the AEC was avoiding the necessity of joining the battle and defending its program. Hence, the SIPI was bringing the agency into court.

Citing the possibility of illegal diversion of plutonium in a breeder-plutonium-fuel economy for paramilitary or terrorist uses, the SIPI statement recited alternative methods of producing power, which would not expose the public to the hazards of plutonium. These included the solar- and geothermal-energy sources which I have discussed, plus more efficient and cleaner methods of burning coal and oil. It was of the greatest importance, the statement added, that the environmental effects of the breeder be fully considered before the development program acquired an "irreversible momentum."

Tweedledum and Tweedledee

Three weeks after the SIPI suit was filed, the AEC issued a Draft Environmental Statement on the LMFBR demonstration plant that it proposed to build in partnership with industry. The agency took the position that it was not required by the NEPA to cover the entire program or to list alternatives, as the SIPI suit demanded, but merely to discuss each installation separately, in its turn. This position created a specific issue for the court.

The demonstration plant, said the AEC, would be a 300- to 500-megawatt, electrical, sodium-cooled, fast-neutron reactor, fueled by a mix of plutonium oxide and uranium oxide. It would be owned and operated by an electric-utility company which would be required to obtain a construction permit from the AEC and to meet other federal, state, and local regulations. Thus, the agency would sit in judgment on the licensing of a new reactor type which it was developing in partnership with industry, the applicant for the license.

The arrangement might be called the Tweedledum-Tweedledee reaction. Tweedledum would submit a Preliminary Safety Analysis Report and an Environmental Report, in accordance with Tweedledee's Regulation 10, CFR Part 50, Appendix D. Upon this action, Tweedledee would issue an environmental statement assessing all the effects which Tweedledum's enterprise would have on the surrounding air, land, and water. Prior to the issuance of an operating license, Tweedledum must submit a final environmental report which would be followed, after appropriate review and comment by the agents of Tweedledee, with another environmental statement authored by Tweedledee.

The burden of the AEC's Draft Statement was that the environmental impact of the LMFBR demonstration plant would be similar to, but in several aspects less than, that of a light-water nuclear steam electric plant, for the following reasons: first, because of design improvements restricting radioactivity release, less radioactive waste would be emitted to the environment than by existing light-water reactors; and second, because of its greater efficiency, the fast breeder would release less waste heat to the environment than existing light-water reactors. Its efficiency of 40 per cent would equal that of fossil-fuel plants. That seemed to resolve the question of thermal pollution from the agency's viewpoint, while under ordinary operating circumstances there appeared to be no problem of radioactive pollution at all. Or would there be?

"The demonstration plant will be designed so that no routine releases of radioactive effluents to the environment will occur during normal plant operation," the AEC draft statement said. However, the draft admitted, some gaseous release may occur over the lifetime of the plant through diffusion and leakage in or around seals, but the technology exists to restrict such releases to a "very low level during normal operation so that they should not have a significant, adverse effect on the environment and will probably not be measurable at the site boundary."

Even a casual reading of this part of the statement suggested a contradiction between the promise that no routine releases of radioactive effluents would occur and the pledge that leakage or diffusion through seals could be held to a low level during normal operation. The conflict of assurances recalls the legendary plea of the Chicago politician caught with his hand in the till—that he never took the money and besides he would put it back.

In the initial fuel charge and after breeding, the demonstrator would contain from 900 to 1600 kilograms of plutonium, the draft said. No safety or environmental considerations were involved beyond those already considered for the light-water reactors. Existing technology was fully capable of handling, processing, storing, and transporting plutonium without risk, according to the statement. While liquid-sodium coolant "reacts readily" with oxygen and water, the "consequences of sodium fires will be precluded or inhibited by special systems or design features." So far as evaluating alternatives to the demonstrator was concerned, that had been done and had resulted in a "reaffirmation of the need for a 300–500 megawatt, electrical, LMFBR demonstration plant." However, the draft statement did not reveal the data or the logic which led to the "reaffirmation" of faith in the breeder as the instrument of man's future well-being. The statement merely concluded "that the environmental impact associated with the use of nuclear-power plants will be reduced by the introduction of the LMFBR into the energy economy."

Snow Job

In September 1971 Dean E. Abrahamson of the University of Minnesota and Arthur R. Tamplin of the Lawrence Radiation Laboratory issued a commentary on the AEC draft statement in behalf of SIPI. In their judgment, it was a snow job.

They noted that although SIPI had called for an environmental-impact statement on the entire breeder program, the AEC's draft referred to a single demonstrator. In its formal reply to the SIPI complaint, the AEC said that publication of the draft and the final statement which would follow, would render the SIPI case moot. It most certainly would not, said Abrahamson and Tamplin.

They characterized the draft as "extremely superficial and inadequate." It was not only "inaccurate and misleading" in what it said and in what it omitted, they charged, but also it was "worthless."

The two scientists said that they were deeply disturbed to reach such a conclusion. What was the AEC trying to do? Was it trying to hide important facts about the breeder from the public? Did it take the NEPA seriously? the critics demanded.

The public, contended the scientists, should be told: (1) The projected number of commercial LMFBRs and the fuel-reprocessing plants that the new technology would have to provide; (2)

where these would be located; (3) the amount of uranium ore the program would need; (4) the volume of radioactive wastes and plutonium that would ultimately be produced in the new generation of reactors; (5) how the wastes and the plutonium would be shipped; (6) how the wastes would be stored—and where—in Kansas; and (7) what the problems might be of safeguarding these materials, of preventing hijacking and the development of a plutonium blackmarket.

The public also should be informed, the critics said, of the environmental problems of mining and milling uranium, including the degree of land disturbance, water pollution, and radom exposure of miners; the likelihood of an LMFBR meltdown or explosion that would release radioactive wastes and plutonium to the biosphere; the prospect of accidental releases of radioactive wastes or plutonium during handling, shipment, or storage; the problem of LMFBR thermal pollution (which the AEC dismissed), and the problem of long-distance electricity transmission.

These questions should be considered in the context of a fullblown LMFBR industrial economy and not merely as they might arise if there were only a few breeders. The public should be informed not only of the prospects of fusion, solar energy, and geothermal power, but also of the relative merits or disadvantages of other fast-reactor types, such as the gas-cooled fast reactor or the molten-salt breeder.

Nor were the critics satisfied with a draft reference to the agency's "over twenty-five years of operating experience in safely handling similar [radioactive] materials at the AEC's production and processing facilities." That record, they felt, was not unblemished. Perhaps, the critics suggested with more than a hint of irony, the agency was referring to the experience obtained at its Rocky Flats plutonium facility, where plutonium fires broke out and plutonium leaked into the environment beyond the plant? Perhaps the agency was thinking of its Nevada Test Site, where one underground test explosion after another had vented radioactive gases into the atmosphere. Or perhaps this record took into account the dispersal of radioactive tailings from the uranium mines for use as land fill in Colorado before the AEC realized that the fill was a radioactive hazard to occupants of buildings constructed on it. Or perhaps, the critics added, the persistent reports of leaks of radioactivity from the fuel-reprocessing plant at West Valley, New York, had never been brought to the attention of the author of the "twenty-five years of experience" section of the AEC Draft Statement.

The draft statement was also criticized for its promise that high-level waste would be treated in accordance with the AEC's storage policy. "It would be of substantial interest," they commented, "to know what this policy is, how it is being implemented, and the environmental impact of that policy."

The critics said it was apparent from the draft statement that the LMFBR was selected over other breeder types because more effort had been devoted to developing it. The overriding consideration seemed to be the need for an early development date. What was the hurry? If industry was pushing, why didn't industry put up the development money? The AEC's predictions of increased power demand were based on the projections made by utility companies, and the draft statement offered no convincing evidence that these projections represented real needs.

The critics noted that the revised AEC criteria for reducing radionuclide emissions from light-water reactors do not, under existing regulations, apply either to the LMFBR or to fuel-processing plants. In the absence of such a requirement, they said, the waste-control sections of the draft simply have "no credibility." In addition, the draft statement was vague about the possibility of plutonium in the fuel cycle reaching critical mass. This point, they said, "cannot be dismissed in two sentences containing only vague and inconclusive statements." The probabilities should be "honestly assessed." So should the consequences of an accident. But no attempt is made in the draft to assess the possibility and consequences of a fuel meltdown. "Until the possible accidents are described and the factors that might lead to them discussed, the public has no basis upon which to form an intelligent assessment of the risks posed by the LMFBR demonstration plant."

In January 1972, the AEC announced its decision on the first demonstration breeder reactor. It would be a 300- to 500-megawatt plant, to be developed jointly by the Commonwealth Edison Company of Chicago and the Tennessee Valley Authority. Cost was estimated at $500 million, with the government picking up half the tab. Conditional pledges of $240 million to be paid over ten years had been made by privately, publicly, and cooperatively owned utilities, according to the AEC.

The deal had been worked out with the secrecy of a military operation. Very little of its details was leaked to the industry in advance—or by the industry during negotiations in the fall of 1971. Commonwealth Edison would provide project management and the TVA would take charge of construction. The reactor would be built by a consortium of suppliers, including Atomics International Division of North American Rockwell, the General Electric Company, and the Westinghouse Electric Company. An entity called the Breeder Reactor Corporation, consisting of representatives of the utilities, the Edison Electric Institute, the American Public Power Association, and the National Rural Electric Cooperative Association, was unveiled in May 1972. Its function was to provide "senior counsel" to the breeder project, to serve as liaison with the utility industry, and to disseminate information about the project "on a broad basis."

The location of the plant was not disclosed in the AEC announcement. During negotiations with the industry, the government had indicated a preference for a region northeast of Knoxville, Tennessee, on TVA land.

So went the Battle of the Breeder, in which the main issue was the degree of the AEC's responsiveness to environmental policy, as expressed in the NEPA. Meanwhile, the United States Court of Appeals made its views known on the obligations of the AEC under the NEPA and the result shook the whole Atomic-Industrial Establishment.

10
Milestone at Calvert Cliffs

By the end of 1970, the rising tide of environmental concern in the United States had become an effective restraint on the plans of the Atomic Industrial Establishment to expand nuclear-power generation and use bomb technology to increase the recovery of gas and oil. The pace of nuclear-reactor construction had been slowed dramatically by citizen intervention on the grounds of safety and pollution at various stages of the AEC's licensing procedure.

Industry spokesmen were alarmed. They complained that the licensing hearings for construction and operating permits had become adversary proceedings. There, the people were arrayed against the electric utilities and, often, against the AEC itself. As I have related earlier, the effect of this widespread citizen intervention, backed by the expertise of dissident AEC scientists, had persuaded the agency to reduce by a hundredfold the radionuclide-emission levels recommended as permissible by the ICRP. The AEC was still plastic enough to respond to public pressure, but only in a limited way. Essentially the AEC was dominated by the demands of the industry and, even though an executive agency, it was subservient to the will of the proindustry Joint Committee.

Intervenors had sought to use the safety issues as levers with which to raise objections to thermal pollution, but the AEC had refused to consider the issue. It had contended that its writ did not run to the question of thermal pollution and in 1969 the agency was sustained by the courts. Until 1970, the statutory basis for issuing a construction permit was the consideration of the "common defense and security" and the "health and safety of public."

The AEC interpreted this language to mean that only the safety issue could be raised at the construction- or operating-license hearing.

On January 1, 1970, this situation changed. The National Environmental Policy Act became effective. It required all federal agencies with licensing authority to consider the environmental consequences in licensing any facility or activity. When the AEC promulgated regulations, under pressure from environmentalists, to implement the NEPA and the new Water Quality Improvement Act, it appeared to the environmentalists that the agency was trying to circumvent the Act—not to implement it. The regulations attempted to minimize the potential delay of considering environmental matters in a license proceeding by providing that if a federal, state, or regional agency had established certain environmental standards, "the only issue to be decided at the hearing is whether the facility will comply with those standards." [1] In this way, the AEC shed its responsibilities for all aspects of environmental pollution except radiation. The AEC regarded radiation standards as a federal matter which it was peculiarly qualified to determine, irrespective of any contrary standards by the states. However, in all aspects of chemical and thermal pollution, certification by any other governmental unit that its standards had been met would dispose of the issue, according to the new regulations. "This conclusion is based on the uncertain premise that the AEC's regulations will stand up under challenge in the courts," observed the Administrative Conference of the United States.[2] It was a prophetic qualification.

The persistent intervention of citizens in nuclear-power-plant licensing hearings had retarded the whole licensing process to the point where Representatives Holifield and Hosmer complained in a letter to Seaborg dated September 24, 1970:

> We have been observing with dismay what appear to be indications of serious deficiencies in AEC's procedural and administrative mechanism for the licensing of nuclear-power plants. A few years ago, a seven to nine months interval was the normal span between the submittal of an application for a construction permit. Now the

processing time is closer to eighteen months and approaching twenty-four months in some cases. . . . Orderly, beneficial licensing procedure appears to be degenerating and perhaps even to have broken down. In contested cases, all semblance of order seems to be disappearing.³

One of the concerns of the two California congressmen was an industry complaint that in contested hearings, "large numbers of management and technical personnel have to waste hundreds of hours attending Board hearings in order to be on hand when, from time to time, safety or related technical issues are heard."

In Hosmer's view, reactor licensing was in a "mess." So he had described it in a speech to the New York Metropolitan Section of the American Nuclear Society on May 11, 1971. And matters were likely to get worse unless the rules on intervention by the public were changed. The AEC had been trying to "unplug the constipated licensing machinery," he said, "but I do not think the Commission plans to crack down nearly hard enough on intervenors bent on making a mockery out of the licensing process." Hosmer characterized the AEC's licensing procedures under the 1954 Atomic Energy Act as "a red-carpet invitation to intervenors, in particular those wily ones whose goal is to prolong the proceedings or completely obstruct the licensing process by utilizing every litigation technique in the hope of eventually uncovering trivia."

Holifield and Hosmer wanted to change the licensing procedure to restrict public intervention in license hearings. That would remove or at least reduce the most serious roadblock to new plant construction. "We are still reeling under the ambiguous impact of several new environmental laws," Hosmer complained. "We are faced with a problem which threatens this nation's electrical energy supply."

Buffeted as they were by "ambiguous impact," the California congressmen stood firm in the conviction that environmental-protection laws inhibited the fulfillment of the critical need for power development. Holifield asserted:

> I have repeatedly warned of power shortages, as have been evidenced by blackouts and brownouts, and I have called for our na-

tion's electric utilities and appropriate agencies of government at all levels to redouble their efforts and join forces to establish realistic, long-range plans for the selection and utilization of sites for large generating facilities, so as to best meet the dual public demand for electric power and environmental protection.[4]

It was true that the public demand for environmental protection was growing; but it was the utility industry that was stimulating public demand for electricity, by pushing sales of appliances and by setting promotional rates that decrease with increased service, so that the customer is encouraged to use more power. A residential consumer of TVA, for example, pays 4 cents for the first 75 kilowatt hours (KWH) a month, but only 1 cent per KWH if he uses more than 500 KWH during the month.[5]

AEC Strikes Back

Beyond the windy rhetoric of the Joint Committee hearing room in the Capitol, the power industry, facing embattled citizens before licensing boards, found it was becoming increasingly vulnerable to assault. A new crop of lawyers, young, aggressive, persistent, and clever, were specializing in environmental law—a field that would attract both idealists and opportunists who were smart and tough. They sold their energies and talents to environmental groups, ranging from long-established national organizations to *ad hoc* committees, and were giving the hearing boards and the utilities a hard time, constantly challenging board procedures that led to long and costly litigation of environmental and safety issues.

In California, Colorado, Minnesota, Wisconsin, Michigan, Illinois, Pennsylvania, New York, all over New England, Maryland and Florida, utilities were struggling with intervenors, sometimes winning, often seeking a compromise settlement of the issues. The utility industry and its allies in the supply field turned to the AEC and the Joint Committee for relief. In the late spring and early summer of 1971, the AEC and the JCAE moved to blunt the legal weapons of the intervenors by restricting the opportunity for intervening in a license case.

Inasmuch as the utilities were most vulnerable to citizen intervention at the stage of the operating-license hearing—after the plant had been built—the AEC proposed to amend its licensing procedures to shut out public interference at that stage. The tactic was so elementary that observers wondered why it had taken the agency so long to figure it out.

Why Have Hearings?

Under the Atomic Energy Act of 1954, the Atomic Safety and Licensing Boards are required to hold public hearings on a utility's application to construct and also to operate a nuclear-power plant. In Minnesota and Michigan, intervenors had been able to exert strong pressure on utilities to reduce radioactive and thermal pollution at the operating-permit stage by threatening expensive legal delays in starting up the plants. The entire industry was worried about its vulnerability to intervention by the public at these stages of the licensing procedure and the industry had the sympathy of the both the AEC and of some members of the Joint Committee. A handful of environmentalists and their lawyers could hold up development of huge power programs under these regulations. "Back in the old days before intervenors," Representative Hosmer reminisced, "we had a press conference to give the natives a look. The natives usually didn't show up. Then the people from Mills Tower [San Francisco]—the Sierra Club—they began to show up. Why do we have to have hearings? Why don't we make it discretionary?" [6]

To correct the industry's Achilles heel, the AEC drew up two amendments to the licensing process for the consideration of the Joint Committee. One (H.R. 9285-S2151) would speed up licensing by removing a mandatory review of each application for a reactor license by the Advisory Committee on Reactor Safeguards (ACRS). No review of the application would be made at all unless either the AEC or the ACRS demanded it. That lifted one source of interference. The second amendment (H.R. 9286-S2152) limited public intervention to a mandatory hearing by the licensing board on the suitability of the reactor site. Intervention was allowed at the construction-permit stage provided that the intervenor could raise "an unresolved question significantly affecting the health and safety of the public." Since such questions certainly would have been resolved at the site hearing, the construction-hearing door was closed. In any case, the amendment provided that the AEC could issue the construction permit at the site-selection stage, thus obviating the possibility of any further public intervention in the licensing process.

Commissioner Ramey explained: "The necessity for preventing unnecessary delays in the construction of vitally needed electric-generating facilities justifies placing this requirement on persons seeking an opportunity to be heard on the details of the plant design after the basic questions of site suitability have been publicly considered." [7] Ramey regarded the existing requirement in the Act—that a hearing be afforded if requested at the operating-license stage—"as even more out of step with the obvious need for early public participation in the regulatory process."

During a hearing on the proposed licensing amendments on June 22, 1971, before the JCAE's Subcommittee on Legislation, Representative Holifield asked AEC Chairman Seaborg about the state of public acceptance of nuclear reactors.

"I don't know if anyone can answer that," Seaborg replied. "We do know we have a number of vocal critics, but I believe they do not represent the majority."

Ramey told the subcommittee that the average time for issuing construction permits had increased from over ten months in 1967 to more than twenty months in 1971. Part of the delay, he noted, was caused by the intrusion of environmental considerations under the Environmental Protection Act and the 1970 Water Quality Improvement Act.

However, he added, delays in some cases were a result of "complex legal questions arising for the first time—for example, the problem of coping with the difficult question of privileged information [in the Monticello case]—and the sharp joinder of environmental and safety issues by a new breed of intervenors with aggressive trial counsel."

Ramey said the reduction of opportunities for public intervention was justified in the light of the AEC's mission of "providing sufficient energy to meet the requirements of our society." Whether that was the AEC's responsibility was moot, but in this context the agency and members of the Joint Committee expressed the conviction they were trying to save the American public from itself, or from the wiles and wicked purposes of the environmentalists. From the point of view of the Atomic Industrial Establishment, time was of the essence. Nuclear power had come just in time to save the country, and there were those people trying to prevent it!

"Does any other process in our society allow for so much public intervention on technical matters?" demanded Representative Holifield. "Any irresponsible person can go before a hearing and make a wild statement and the press will print it, while ignoring the trained technicians." Holifield complained that two television documentaries on nuclear power for which he had been interviewed had been distorted into "diabolical misrepresentations" by mass media tactics "seeking to scare the people. This is why you have abroad in the land fear and apprehension," he said. "I will never again go on a TV program. I've been burned twice."

The New Regulators

The utility industry sent its spokesmen to the hearings with a basket of complaints, but no unanimity on whether the amendments the AEC proposed would eradicate public interference. The industry wanted to get rid of a good deal of government interference, too. It wanted to build nuclear-power plants without anybody butting in.

John G. Quale, president of the Wisconsin Electric Power Company and its subsidiary Wisconsin-Michigan Power Company, told his story. The utilities are joint owners of the Point Beach Nuclear Power Plant at Two Creeks, Wisconsin, consisting of two pressurized-water reactors, each with an electrical output of 497 megawatts.

Quale said the AEC proposal did not go far enough. It would not solve the problem "resulting from the use of the Commission's hearing procedures for ulterior purposes, including that of extracting concessions from applicants on matters which may or may not even be related to the legitimate issues of the proceeding."

As a case in point, Quale cited the experience of his companies' Point Beach Nuclear Plant Unit Two. The Wisconsin Ecological Society and the Wisconsin Resources Conservation Council intervened at the operating-license stage of Unit One. The intervenors were concerned about radioactive emissions.

"After considerably difficult negotiations," Quale related, it was agreed that the utility would install additional radioactive-waste-control facilities, which would reduce radioactive emissions as far as the state of the art would allow, below AEC limits. The intervenors then withdrew their complaint.

But the utilities' troubles were not over. No sooner did the AEC post notice of a proposed issuance of an operating license for Unit Two than a petition to intervene was filed by three other groups—the Businessmen for the Public Interest, the Sierra Club, and a group called Protect Our Wisconsin Environmental Resources (POWER), which Quale said was "apparently organized only for purposes of intervention in our case."

Quale complained that it was clear that the intervenors were going to conduct a *de novo* analysis of the plant and its design, and then decide whether in their judgment the design was okay:

> The intervenors have thus set themselves up as a quasi-regulatory agency, responsible to no one, with the expectation of a Commission-countenanced authority to conduct virtually unlimited fishing expeditions into the activities, books, records, and thought processes of the applicants, the contractors, and the Commission staff.

Quale thought it was clear that the purpose of the intervention in the Point Beach Unit Two hearing was not to resolve nuclear-safety questions, "but rather to use the complexities of the nuclear hearing process to coerce the applicants to install closed-cycle cooling facilities acceptable to the intervenors."

Closed-cycle cooling, usually requiring the construction of cooling towers, is considerably more expensive than simply drawing water out of a river or lake to cool the nuclear-power-plant condenser and then discharging the heated water back into its source. The method of coercion, he said, was to "impose sufficient expenses and operating difficulties upon the utility by delays in the licensing procedure as to make it expedient to yield." As an example of yielding, he cited the agreement involving the Palisades plant of Consumers Power Company between the utility and intervenors.

Another witness, Thomas G. Ayers, president of the Commonwealth Edison Company of Chicago and chairman of the Edison Electric Institute's Policy Committee on Atomic Power, confirmed that there was rising concern in the industry about licensing delays. He spoke as if a crisis was at hand:

> Compounding a difficult situation is the growing number of contested licensing proceedings for nuclear-power plants. We are most concerned over the rising number of contested proceedings for operating licenses, which cause nearly completed generating plants to lie idle while power may be in short supply and costs pile up.

Ayers knew about costs piling up from bitter experience. Each week that his utility's Dresden II at Morris, Illinois, was kept from receiving its operating permit, had cost the utility $600,000, he said.

In addition to the AEC and the industry, there was a third force at the subcommittee hearing on licensing changes—the intervenors themselves. They had a word or two to say about the AEC's attempt to limit their intervention. One spokesman was Mrs. Carl of the Lloyd Harbor Study Group, mentioned in Chapter 5—the group that had been engaged for fourteen months in hearings on the construction permit for the Shoreham Nuclear Power Station of the Long Island Lighting Company. Mrs. Carl contended that the hearing procedure, even as it was, discriminated against citizens, in favor of industry and the AEC. She charged that expert testimony of forty witnesses assembled by the intervenors had been either excluded from the hearing record on procedural technicalities or cut and deleted as it was presented.

"Are these hearings truly information-seeking?" she demanded. "Or are they only industry protective?"

The veteran radiation critic Dr. Sternglass contributed a comment on the proceedings. It was vital, he said, that there should be the widest possible opportunity for specialists to participate in the examination of the safety of nuclear reactors because, "time and again in the history of science, unexpected and upsetting discoveries have often been made by outsiders."

It was the public's concern whether a nuclear-power plant should be constructed and operated, as well as the concern of the AEC, other witnesses insisted. Harry Blanchard, a high-school teacher of chemistry, physics, and mathematics, representing an eastern Pennsylvania ecology group, testified that the elimination of the mandatory public hearing at the construction- and operation-licensing stages would effectively shut the public out of the decision-making process.

Mrs. Daniels of the Wyoming County Citizens Committee for Environmental Concern said her organization regarded the proposed licensing changes as a means of speeding the construction of nuclear reactors "with the least possible interference from the public."

Foiled by the Voters

The strategy of the AEC and the industry to reduce environmentalist "interference" in the licensing process was doomed from the start. First, the Atomic Industrial Establishment was not unified on a means of tuning out the "static" from the public. Second, the inflated rhetoric of the agency and of some members of the Joint Committee seeking to justify the exclusion of the public from critical stages of the licensing process served only to confirm the antienvironmentalist posture of these bodies. The AEC gave lip service to mounting concern about the environment, but now displayed bureaucratic impatience and brusqueness toward concerned organizations.

The net effect of the proposed amendments to the licensing process was to clarify the agency's public-be-damned image and polarize the attitudes of the Atomic Industrial Establishment on one hand and of the environmental organizations on the other. Politically this created an intolerable situation for an Administration which was going to need all the good will it could find to win reelection. Least of all could the Administration afford to alienate the older environmental groups with a predominantly conservative membership.

The AEC-industry cant about the imminence of an electric-power crisis simply had no impact. The public was unconcerned, probably because it doubted the credibility of the government-industry partnership. On the other hand, ten years of rising concern about the deterioration of the environment as a result of governmental indifference to the control of industrial wastes had alarmed an active, articulate segment of American society.

The big mistake of the Atomic Industrial Establishment was to underestimate the political power of this segment by trying to exclude it from the exercise of its rights in the critical stage of the licensing process. The amendments had little support on Capitol Hill and a great deal of opposition.

By the end of the summer, the proposed amendments had been dumped. Representative Holifield announced that in the light of other developments affecting the licensing of nuclear-power plants, the subcommittee had not deemed it advisable to report the amendments for consideration by the full Joint Committee.

Calvert Cliffs

Other developments arose out of the construction of a 1600-electrical-megawatt nuclear-power plant by the Baltimore Gas and Electric Company on the western shore of Chesapeake Bay near Lusby, Maryland, about thirty miles from Washington, D.C. The plant, called the Calvert Cliffs Nuclear Generating Station, consisted of two pressurized-water reactors manufactured by the Combustion Engineering Company, each with an output of 2450 thermal or 800 electrical megawatts. Starting with site preparation in 1967, the utility planned to put the first unit into operation in January 1973, and the second unit into service a year later.

During hearings on the construction permit in 1969, a group of Maryland citizens, most of them residing between Washington and the Bay, intervened to raise questions about the radiation and thermal pollution of the great estuary from the first nuclear-power plant to be erected on its vast shoreline. The citizens had first become concerned about the project when the utility disclosed it would have to cut a 47-mile swath through the forests and fields of the Maryland countryside for power-distribution lines from the plant. As they learned more about it, the focus of their concern shifted to the pollution of the Bay, the greatest natural resource of the state.

The citizens formed the Chesapeake Environmental Protection Association (CEPA), enrolled experts in its cause, and intervened in the construction hearing. One of CEPA's key witnesses on the radiation-pollution issue was Dr. Edward P. Radford, professor of Environmental Medicine at The Johns Hopkins School of Hygiene and Public Health in Baltimore. He told the hearing examiners that tritium released from the proposed plant could be a health hazard to a significant population who ate seafood from the Bay.

Later, in testimony before the Joint Committee on January 28, 1970, Radford explained that tritium is probably the most important radioactive waste from nuclear fission, chiefly because it is difficult to control at fuel-reprocessing plants. At the General Electric fuel-reprocessing plant at West Valley, New York, he said, a small creek running from the plant had an average of 700 picocuries per milliliter of tritium, with occasional values of 1000 picocuries per milliliter. These are relatively high values. On the assumption of a straight-line relationship between rate of decay and mutation probability, he said, "the present maximum permissible concentration levels, if continuously ingested, would lead to

3300 new mutations per million human births, or about 12,000 mutations per year in the United States."

At this point, Representative Holifield, then JCAE chairman, interposed.

"There," said he, "is your headline."

During the earlier licensing-board hearings on the Calvert Cliffs construction permit, the utility had countered that the tritium issue was irrelevant, inasmuch as it attacked the prescribed radiation standards in the Code of Federal Regulations, Part 20, a passage generally regarded by the Establishment as gospel.

This position was upheld by the AEC's legal staff, even after a Review Board had expressed interest in its merit. As to the issue of thermal pollution from the plant, the AEC again took the position that this concern was beyond its purview and hence was irrelevant.

In his appearance before the Joint Committee, Radford recalled this bit of licensing history. He asked:

> If the adequacy of the standards by which the AEC permits release of radionuclides cannot be challenged at these public hearings and if the issues of thermal pollution, power-line location, hazards from fuel reprocessing, and other matters cannot be raised, what scientific basis is there for public discussion?

There was no answer. The question had been only rhetorical so far as the AEC was concerned, for the agency had issued a memorandum dated August 8, 1969, stating in part:

> Further, it should be clear that our licensing regulations, which are general in their application and which are considered and adopted in public rule-making proceedings, wherein the Commission can draw on the views of all interested persons, are not subject to amendment by boards in individual adjudicatory proceedings.[8]

From a layman's point of view, Radford observed, these statements precluded any substantive challenge to the radiation standards. The impression of the members of CEPA was that the AEC would allow them to debate nothing in the realm of radiation or thermal pollution. Thus, from the viewpoint of the intervenors, even the public hearing was stacked in favor of the utility.

In Hot Water

Like similar organizations in Pennsylvania, New York, Michigan, Wisconsin, and Minnesota, CEPA was an *ad hoc* group, formed specifically to do something about the thermal pollution and radiation hazard of a nuclear-power plant.

One of its founders, a retired engineer named Y. Kirkpatrick-Howat, said the group suspected that the thermal pollution of the Bay would be appreciable: first, because the plant was nuclear and therefore less efficient than a fossil-fuel plant; and, second, because it was located at the narrowest point of the Bay, where the estuary was only six miles wide, compared to its average width of twenty miles. The hot-water plume discharged by the plant's cooling system would extend all the way across the Bay, forcing fish and other organisms which spawn in the Upper Bay to run a gauntlet of hot water.

The hot-water plume would be a product of a type of "once-through" cooling system, which Baltimore Gas and Electric had decided to use to cool the plant condensers.

All steam-electric plants require a large flow of water through their condensers to convert exhaust steam from the turbine into water, so that it can be recirculated back to the boiler. In passing through the condenser, the coolant water is heated 10 to 20 degrees Fahrenheit or more, depending on plant design. Waste heat picked up by the coolant water has to be dissipated to the environment in some manner. The cheapest way is simply to dump the heated "coolant" back into the same place it was taken from—a river, lake, estuary, or ocean.

Baltimore Gas and Electric planned to pump 5400 cubic feet per second of Bay water through the unit condensers and discharge it offshore near the surface. It specified that the maximum temperature rise in the water as it passed through the condenser would be 10 degrees Fahrenheit.[9] According to its own thermal-effects studies, the utility contended, this rise in the temperature of the exhaust water would have no deleterious effect on the fish.

CEPA didn't buy the utility's evaluation.

"There's this little 'possum fish and he migrates up and down the Bay, day and night, to keep his temperature in equilibrium," said Charles R. (Sonny) Tucker, a CEPA organizer who operates a farm near Harwood, Maryland. From Tucker's farm, a powerline swath through the woods is plainly visible, a reminder of the environmental effects of electric power. "This fish is extremely susceptible to temperature change," said Tucker. "So when he hits a thermal plume, he may go into shock and die, or else he goes crazy and swims back downstream, instead of up."

Tucker and other members of CEPA said that, aside from a few species known to be temperature sensitive, the impact of thermal pollution on the marine populations in the Bay was a massive "unknown." The citizens didn't know what it would be and neither did the utility. Most certainly, the AEC didn't know and the agency had demonstrated that it didn't care.

The problem was: nobody knew much about the effects of thermal pollution. And the AEC, which was licensing a thermal polluter, had no intention of finding out. Were the citizens supposed to undertake this responsibility? Tucker wondered.

Howat, the retired engineer, presented his view of the issue:

> Philosophically our position has been that we're not opposed to progress in any way, shape, or form. But we're violently opposed to making major environmental changes without having established proper base-line studies and without having, so far as human capabilities can stand, attempted to determine what the end effects will be. I am opposed to the AEC being charged with two incompatible tasks, one to promote the peaceful use of atomic energy and the other to be the sole judge of the effects thereof. This is absolutely ridiculous.

In spite of the lack of specific and detailed studies of the effects of heated water on the biota of the Bay, it was generally known that temperature is critical in the physiological processes of marine organisms, according to a Federal Power Commission staff study.[10] An increase in the temperature of water reduces its capacity to hold oxygen and increases the rate of biological activity.

The FPC study said that, although few investigations have been made covering whole biological communities,

> it is known that temperatures higher than those normally experienced, particularly during summer months, can be detrimental to aquatic organisms in a variety of ways: it may make them more susceptible to disease or to the poisoning of their food supply; their ability to catch food may diminish; the inability to reproduce or compete with other aquatic organisms may eliminate a whole population in a subtle way.

In addition, the elimination of one species in the food chain may change the ecological balance and cause significant changes in species dependent on the chain.

"All aquatic species have an optimal temperature range," the study noted. "If the temperature varies above or below this range, the chances of survival for that species incrementally decreases."

A Bereaved Giant

One of the most outspoken critics of the Calvert Cliffs nuclear plant was Jess W. Malcolm, executive director of the Chesapeake Bay Foundation, Inc., of Annapolis, a charitable and educational organization founded in 1966 to protect the estuary from polluters.

During 1969 and 1970, Malcolm waged a vigorous campaign, almost single-handed, to obtain independent studies of the effects of the plant's thermal plume on the biota of the Bay. In one report, he wrote:

> Such an evaluation has not yet been made and, by virtue of the main unanswered questions still being asked, the Baltimore Gas and Electric Company finds itself in the role of a bereaved giant, annoyed by the persistent buzzing of a swarm of gnats attracted to its sweaty brow.

Malcolm said that he met with company officials in 1968 in an attempt to persuade them to cooperate in a study by an independent review committee—one that would not be on the uitility's payroll. But the company would not agree, "preferring instead to run the gamut of growing public indignation."

In the midst of his spirited contest with the company, Malcolm was replaced as director by Arthur W. Sherwood, one-time unsuccessful candidate for Mayor of Baltimore. Sherwood adopted a more neutralist attitude toward the Calvert Cliffs controversy and challenges to the utility by the foundation stopped.

From the viewpoint of its own studies and convictions, the utility directorate had every right to feel aggrieved. The once-through cooling method planned for the Calvert Cliffs plant was as old as fossil-fuel steam electric plants. Thermal pollution was simply a slogan of an incomprehensible, antiutility movement which apparently sought a return to the days of candles and muscle power.

Was thermal pollution actually pollution? Environmentalists believed so, because it altered the environment in ways which might be inimical to life. With the advent of the light-water nuclear reactors, the thermal effects of the once-through method became aggravated. The light-water reactor types coming into use in the United States required 50 per cent more cooling water through their condensers than modern fossil-fuel plants. Light-water reactors are considerably less efficient than fossil-fuel plants, producing more waste heat in relation to the amount of electric power they generate. Consequently the proliferation of nuclear-power plants on the rivers, lakes, and estuaries of the land posed a significant threat to biota in those waters, in the environmentalist view.

Did this mean nuclear plants should be ruled out because of thermal pollution? Not at all. There were other methods of cooling the plant condenser which did not discharge hot water in rivers, lakes, or estuaries—closed-cycle methods. In Great Britain, cooling towers in which heated water from the condenser is cooled by evaporation had been developed, largely because sizable inland bodies of water were not available for cooling near the plant sites. Cooling towers were springing up in the United States also, partly as a result of environmentalist pressure. That, as we have seen, had been the outcome in Michigan where powerful environmental groups had forced the Consumers Power Company to agree to install cooling towers and reduce radiation emissions from the Palisades plant, under the threat of law suits.

In the Evaporative Cooling Tower, hot water from the condenser is passed through the tower where the temperature is lowered by evaporation. The water is then recirculated to the condensers. The evaporation may be accelerated by fans, but the flow of air is also generated by natural draft through the chimneylike structure. Natural draft towers are big, hyperbolic structures, about 400 feet in diameter and 400 or more feet high.

According to Federal Power Commission studies, cooling towers cost more and use more water than the once-through method. They are also likely to create artificial fogging and icing problems locally.

AEC and NEPA

On the theory that its writ did not run beyond radiological pollution, the AEC rejected the intervenors' protests against thermal pollution, as in the Calvert Cliffs case. What concessions to thermal pollution had been made in Michigan and in Wisconsin had come about as a result of "citizen regulation"—not through any effort by the AEC. The Commission did, of course, take cognizance of the National Environmental Policy Act. It announced that its responsibilities under the Act would have to be determined through a detailed (and lengthy) rule-making process. Environmental groups interpreted this as a stall.

In the Calvert Cliffs project, where the thermal-pollution issue was paramount, the AEC had ignored the matter in issuing the construction permit to Baltimore Gas and Electric Company. Concerned citizens' groups had turned to the Maryland Public Service Commission, which in 1968 had been given authority by the state legislature to approve all sites for electrical generating plants and to hold public hearings. But BG&E took the position that its Calvert Cliffs plant was not subject to PSC site authorization, because construction had begun in 1967, with site excavation, before the effective date of the new law, July 1, 1968. The question went up to the Maryland Court of Appeals, which held that the PSC did have jurisdiction over the Calvert Cliffs site and thus could require an assessment of the plant's environmental impact.

In the meantime, Maryland intervenors banded together in a new organization called the Calvert Cliffs Coordinating Committee, or "Quad C" for short. With financial help from the Sierra Club and the National Wildlife Federation, they employed Washington counsel to challenge the validity of the licensing regulations under which the AEC had given the construction permit to Calvert Cliffs.

All three organizations joined in a demand that the AEC order BG&E to show cause why the construction permit should not be suspended, pending a complete study of the environmental effects of the plant. They demanded that environmental statements be issued by the company and the AEC, in conformity with NEPA.

The AEC fell back to its first line of defense. It responded that no decision on these requests would be made until the agency had worked out the rules under which it would implement NEPA. The agency's rule-making procedure, which had been launched April 2, 1970, was still going on in November, with no end visible to the environmentalists.

On November 12, 1970, the petitioners asked the AEC to act on their request in ten days. They pointed out that the Maryland Court of Appeals had ordered a complete review of all aspects of the plant, including its environmental effects, by the Maryland PSC. It would be more logical from an administrative point of view, the petitioners argued, if both the state and the AEC reviews were conducted at the same time.

Again, the AEC stalled. Two weeks later, the environmentalists filed a petition for review of the AEC's refusal to grant their requests in the United States Court of Appeals for the District of Columbia Circuit. They argued that the longer the environmental-impact review was held up, the less feasible it became to make changes in plant design and operation because of the progress of construction. Thus, the public was being "locked into the Calvert Cliffs plant" as the utility wanted to build it, "without any hope of feasible and environmentally more desirable alternatives being accepted." The failure of the AEC to act on the requests, the petitioners said, "was tantamount to a denial of the requests."

The petition asked the Court of Appeals to require the AEC to issue a show-cause order to BG&E to determine why plant construction should not be suspended during a complete review of environmental factors required by NEPA. The AEC then acted: On December 3, 1970, the AEC adopted rules for implementing NEPA and published them in the Federal Register the next day.

Appendix D

The rules, formulated as Appendix D to Part 50 of the AEC's governing regulations, provided in part that:

1. Environmental factors need not be considered by the hearing board unless affirmatively raised by outside parties or staff members.

2. No party to a hearing could raise nonradiological environmental issues if the hearing notice appeared in the Federal Register before March 4, 1971. This rule exempted Calvert Cliffs, along with other plants in construction, from NEPA.

3. The hearing board is prohibited from conducting an independent evaluation and balancing of certain environmental factors if other responsible agencies—state, regional, or federal—already have certified that their own environmental standards are satisfied.

4. When a construction permit has been issued before the NEPA compliance was required, and when an operating license has yet to be issued, the agency will not formally consider environmental factors or require modifications (backfitting) in the proposed facility until the time of the operating license.

The petitioners attacked these rules in a second complaint to the Court of Appeals, asserting that in the context of the NEPA the rules were clearly a violation of law. They argued that:

——By refusing to allow nonradiological environmental issues to be raised at hearings noticed before March 4, 1971, the AEC was being derelict in carrying out NEPA.

——The dereliction was compounded by the agency's refusal to

consider nonradiological environmental factors at hearings unless someone intervened to raise them.

———By accepting certification of compliance with state, regional, or federal standards on nonradiological pollution, the AEC was illegally relinquishing its duty under the NEPA to consider such factors on its own—since it was solely responsible for the licensing of nuclear-power plants.

———The AEC must freshly examine the environmental impact of plants whose construction was started prior to the passage of NEPA; otherwise the balance in any controversy would be clearly in favor of issuing the operating license when the construction was completed.

In regard to Calvert Cliffs itself, the petitioners called for the application of NEPA standards to the plant. They averred that this required a fresh environmental study on the effects of thermal pollution of Chesapeake Bay and backfitting of the plant if the results of the study made it necessary.

The petitioners asked the court to declare illegal the sections of Appendix D which conflicted with NEPA and to instruct the AEC to promulgate new regulations "consistent with the Act." They asked specifically that the court require the issuance of a show-cause order, leading toward suspension of construction of the plant, with backfitting ordered, if found necessary after a study.

To these allegations and prayers, the AEC responded that the legislative histories of both NEPA and the Water Quality Improvement Act of 1970 made it clear that the AEC is obligated to defer to the expertise of other state and federal antipollution agencies, in regard to the nonradiological effects of nuclear-power facilities. The NEPA, in the AEC's opinion, was basically an expression of policy and of procedural requirements. It was not meant to impose inflexible standards on the various administrative agencies.

In the light of the national energy crisis, the agency said it had acted well within its discretion in limiting consideration of nonradiological environmental factors to hearings noticed after March 4, 1971. Such a period of transition is consistent with

NEPA directives that agency action be "practicable" and "consistent" with other essential considerations of national policy.

The AEC considered NEPA as merely one step in an evolutionary process, the end of which receded into the mists of a distant time. Besides, the agency pointed out, new legislation was being considered to regulate the siting of power plants. The AEC's legal eagles suggested this would do more to advance NEPA's goals than an overly stringent enforcement of NEPA itself.

In its official attitude, the AEC leaned heavily on its assumption of a power crisis; indeed, it seemed to be virtually promoting a belief in such an emergency. But it did not recognize the onset of an environmental crisis. Perhaps this species of tunnel vision was a product of the agency's primary concern for promoting atomic power, against which environmental issues loomed only as obstacles. The formulators of AEC policy, who are lawyers and engineers, were blissfully unaware—or so it seemed—of the spoiler image being created by their indifference and, in some cases, outright hostility toward environmental protection.

Because of the proximity of the Calvert Cliffs plant to Washington, D.C., and because of the persistence of the intervenors, the case had become something of a *cause célèbre* on Capitol Hill. Officials of the Office of Science and Technology, which advises the President, were fully aware of it, as they were bombarded with protests and pleas from the embattled citizens of Maryland. No other reactor controversy in the nation had generated quite so much interest at the seat of government. And none appeared to be so fraught with the political peril of tarring a national administration with the environmental-spoiler brush. The AEC was, after all, an executive agency. Its members were appointed by the President and in theory were responsive to his policies. The prospect of this agency becoming Public Enemy No. 1 in the environmental crusade, which had led so many people to equate environmental protection with the survival of civilization, was not one that any rational political apparatus could tolerate. In this light, the spectacle of the AEC dragging its feet on the environmental issue, particularly one affecting Chesapeake Bay, was especially frightening.

Both the initial petition seeking a show-cause order against the plant and the second petiton asking the court to invalidate Appendix D were consolidated by the court. The issues were at once simple and complex. Appendix D represented one view of the NEPA—a view which if carried to its logical conclusion would make the Act a mere statement of policy. CEPA and its allies took another view, regarding NEPA as a weapon. But, like Excalibur, this weapon needed a special hand to draw it from the stone of federal agency indifference, and the hand might be proferred by the court. The first requirement for unsheathing the sword was the arrival of an issue which could be appealed to the court, and the Calvert Cliffs case fulfilled that requirement.

The question of how NEPA was to be applied to atomic power plants was of direct concern to utilities all over the country. Several of them intervened in the appeal as *amici curiae,* namely, Consumers Power Company, Duke Power Company, Long Island Lighting Company, Northeast Utilities, Pacific Gas and Electric Company, Rochester Gas and Electric Company, Virginia Electric and Power Company, Indiana and Michigan Electric Company, and Portland General Electric Company.

Serving more than a million customers in the lower peninsula of Michigan, Consumers Power Company prefaced its brief with the claim that the power demand on its system had doubled in the last ten years and was expected to more than double in the next ten. To meet this growth, the company was completing its Palisades plant near South Haven and a two-unit station at Midland—for which a construction permit was pending. The brief stated, "The proceeding entails a review of an industry-wide rule-making proceeding in which the AEC has endeavored to balance important interests of the public in both protection of the environment and assurance of an adequate supply of electric power."

Consumers Power argued that the Commission's adoption of a transition period for implementing some of the NEPA obligations was "a valid exercise of administrative discretion." The Michigan utility attacked the petitioners' contention that the AEC must review existing federal, state, or regional environmental-protection standards as "inconsistent" with both the pattern of environmental regulation created by subsequent federal statutes and legislation proposed by the President. If this contention was accepted by the court, the utility argued, it would "produce chaos in the electric-power industry and severely hamper the industry's ability to meet the nation's electric-power needs. . . ."

The Wright Decision

The court delivered its decision on July 23, 1971. Judge James Skelly Wright found that the AEC's regulations violated the mandate of NEPA which "imposes a substantive duty upon every federal agency to consider the effects of each decision upon the environment and to use all practicable means . . . to avoid environmental degradation." The case was remanded to the AEC "for proceedings consistent with the court's opinion."

Judge Wright's decision was as far-reaching as NEPA itself. In effect, it activated the law. It prescribed precisely the way in which the law must be applied. The NEPA was one of several recently enacted statutes that "attest to the commitment of the government to control at long last the destructive engine of material progress," he noted. "But it remains to be seen whether the promise of this legislation will become a reality. Therein lies the judicial role. . . . Our duty, in short, is to see that important legislative purposes, heralded in the halls of Congress, are not lost or misdirected in the vast hallways of the federal bureaucracy."

He held that the AEC's refusal to review nonradiological effects of nuclear plants, on which other agencies have passed, conflicted with the NEPA mandate, as did the AEC's failure to require hearing-board review of nonradiological environmental effects unless these were raised by outside parties. AEC's refusal to consider such effects at hearings announced before March 4, 1971, and to consider alteration of plants granted construction permits before NEPA became effective but for which operating licenses were still pending—these, too, violated the mandate of NEPA and were "inconsistent with the Commission's duty to fully consider action which will avoid environmental degradation."

Judge Wright predicted that "these cases are only the beginning of what promises to become a flood of new litigation—litigation seeking judicial assistance in protecting our natural environment."

Perhaps the NEPA's greatest importance is to require the AEC and other agencies to consider environmental issues, he said, just as they consider other matters within their mandates:

> NEPA first of all makes environmental protection a part of the mandate of every federal agency and department. The AEC, for example, had continually asserted prior to NEPA that it had not statutory authority to concern itself with the adverse environmental effects of its actions. Now, however, its hands are no longer tied. It is not only permitted but compelled to take environmental values into account.

The Judge then turned his attention to the question of whether the AEC is correct in thinking that its NEPA responsibilities could be discharged outside the hearing process—whether it is enough that environmental data and evaluations merely accompany an application through the review process, but receive no consideration from the hearing board.

"We believe that the Commission's crabbed interpretation of NEPA makes a mockery of the Act," the Judge commented. What purpose was there in a requirement that detailed statements accompany a proposal through the agency review process, he asked, if "accompany" means no more than physical proximity— "mandating" no more than the physical act of passing certain folders and paper, unopened, to reviewing officials along with other folders and papers?

> What possible purpose could there be in requiring the "detailed statement" to be before hearing boards, if the boards are free to ignore entirely the contents of the statement? NEPA was meant to do more than regulate the flow of papers in the federal bureaucracy. The word "accompany" . . . must not be read so narrowly as to make the Act ludicrous.

Next, Judge Wright examined the Commission's reason for delaying consideration of environmental issues until March 4, 1971—fourteen months after the NEPA went into effect. The excuse was that the AEC wanted to provide for an orderly transition period. The decision said:

> Again, the Commission's approach to statutory interpretation is strange indeed—so strange that it seems to reveal a rather thoroughgoing reluctance to meet the NEPA procedural obligations in the agency review process, the state at which deliberation is most open to public examination and subject to the participation of public intervenors. . . . It seems an unfortunate affliction of large organizations to resist new procedures and to envision massive roadblocks to their adoption. Hence, the Commission's talk of the need for an "orderly transition" to the NEPA procedures. It is difficult to credit the Commission's argument that several months were needed to work the consideration of environmental values into its review process.

Judge Wright added:

> In the end, the Commission's long delay seems based upon what it believes to be a pressing national power crisis. Inclusion of environmental issues in the pre-March 4, 1971, hearings might have held up the licensing of some power plants for a time. But the very purpose of NEPA was to tell federal agencies that environmental protection is as much a part of their responsibility as is protection and promotion of the industries they regulate. Whether or not the specter of a national power crisis is as real as the Commission apparently believes, it must not be used to create a blackout of environmental considerations in the agency review process.

Judge Wright stressed the importance of reviewing the environmental consequences of a plant before the operating license is considered:

> Once a facility has been completely constructed, the economic cost of any alteration may be very great. In the language of NEPA, there is likely to be an "irreversible and irretrievable commitment of resources" which will inevitably restrict the Commission's options.

He added that the Commission should consider seriously the requirement of a temporary halt in construction pending its review and the backfitting of technological innovations. No action which might minimize environmental damage may be dismissed out of hand. Of course, final operation of the facility may be delayed thereby. However, he said,

> Some delay is inherent whenever the NEPA consideration is conducted, whether before or at the license proceedings. It is far more consistent with the purposes of the Act to delay operations at a stage where real environmental protection may come about than at a stage where corrective action may be so costly as to be impossible.

No Appeal

So ran, in part, the landmark decision, setting forth the judicial philosophy of environmental protection. It appeared to release NEPA from the fetters of bureaucratic indifference which had threatened to restrain it to mere rhetoric. Had that not been the fate of the Refuse Act of 1899, virtually ignored for more than sixty years, until the conservationist awakening in America? Environmental-protection laws were only as effective as the determination of the citizens to see them enforced.

The impact of the court's decision is not readily calculable. It not only requires federal agencies to act, but sets forth in clear language that the preservation of the environment under NEPA supersedes any judgment by the AEC, or any other agency, about the nation's power needs. It challenged judicially the shibboleth of the Atomic Industrial Establishment that unlimited expansion of electric power was a goal of the highest priority.

The new considerations for issuing a commercial nuclear-power plant license applied to all permits and licenses retroactively since January 1, 1970, when NEPA took effect. The AEC said the decision affected sixty-three pending license applications involving ninety-one nuclear-power plants, and five plants which were completed by the end of 1969 but for which operating licenses were issued later.

At the end of August 1971, the AEC announced it would issue

new regulations implementing the July 23 decision and would not appeal. That decision was not surprising in view of several developments that had occurred earlier in the summer.

After ten years as AEC Chairman under Presidents Eisenhower, Kennedy, and Nixon, Seaborg resigned July 21, 1971—rather suddenly, it seemed to observers. Only the year before, he had accepted a five-year renewal of the chairmanship from Mr. Nixon. It was announced that the fifty-nine-year-old Nobel laureate planned to return to the University of California to resume his research. Seaborg's resignation had come just two days before the Calvert Cliffs decision.

No less surprising than his sudden decision to quit was his replacement, the forty-two-year-old economist James R. Schlesinger, who had been in the Office of Management and Budget as assistant director. Schlesinger had taught economics at the University of Virginia and had directed a nuclear-proliferation study at the RAND Corporation, oldest of the government's think tanks.

In another surprising move, the President appointed William O. Doub, chairman of the Maryland Public Service Commission, to the vacancy on the Commission left by the death, in an airplane crash, of Theos J. Thompson, a physicist. Doub, a thirty-nine-year-old lawyer, had served on the President's Air Quality Advisory Board. He had been the Maryland PSC chairman during that body's court fight to assert jurisdiction over the Calvert Cliffs plant site. These positions seemed to give him credentials acceptable to the environmentalists.

The implications of the Schlesinger and Doub appointments were so obvious that even the Atomic Industrial Forum's cautious but usually authoritative Newsletter "Nuclear Industry" commented on them as follows:

> Washington observers see the nominations of Schlesinger and Doub, neither of whom has been closely associated with the nuclear community, as an attempt by the Administration to give the Commission a better image in the eyes of the nation as it continues to spearhead the development of nuclear power. . . . [Doub had] won the respect of environmentalists in his handling of the Maryland

PSC hearings on the Baltimore Gas and Electric Company's Calvert Cliffs plant.

With these appointments and Seaborg's departure, the AEC was without a scientist member. Representative Holifield was quoted by *Business Week* as saying that was not a problem, because: "The greatest part of the scientific work has been done. Now, the preponderant task is engineering and the application of the scientific knowledge we already have." *Business Week* ventured the opinion that President Nixon wanted a more dynamic commission that could sell reactor programs to Congress and to the public. A vocal AEC scientific leader, a veteran of the Manhattan Project, commented: "Now, the economists are running the show; next, the environmentalists will."

The Changing of the Guard at the AEC was exactly what it appeared to be: an effort to improve the image, reduce the agency's rather conspicuous credibility gap, and appease the environmentalist movement at the expense of the electrical industry, the gas industry, and the big oil companies that were running much of the energy establishment.

The new appointments also represented a shift in control which brought the AEC more closely under the wing of the White House and made its policies somewhat more responsive to the ideas of the bright young men in the Office of Science and Technology, the President's private think tank. A number of observers felt that the President was moving the AEC out of the orbit of the Joint Committee, which functioned like a board of directors in the Atomic Industrial Establishment. Whether this was a motive for the change may be moot. But that was one result of the new appointments.

The Power Hucksters

Under its legal mandate to develop peaceful uses of atomic energy, the Joint Committee not only funded but gave Congressional approval to a power-expansion policy which it firmly believed was of critical importance to the nation. This belief, of course, was shared by the power industry and its suppliers. So long as it predominated in the Establishment, there seemed to be no way of balancing power development with environmentalist doctrines of conserving energy resources and protecting the environment from radioactive, thermal, and chemical pollution.

Energy Demand had become a cliché which the energy companies exploited to rationalize the plundering of natural resources. It displayed the Topsy syndrome—it just grew. It grew in response to promotional practices of government and industry in encouraging the use of energy by the people. Energy growth was a public good, equated with the expansion of the gross national product and higher standards of living. Growth itself was a good. Its antithesis was decay. As S. David Freeman of the Office of Science and Technology predicted, energy consumption from 1970 to 2000 will be almost three times that of the previous thirty years.[11] Why? No one really knew, since such growth could not be explained in terms of increasing population or in terms of easing poverty. In the absence of a national energy policy, energy supply simply responded to demand as long as the fuel held out, and the power companies kept stimulating demand.

Various state and national agencies had varying degrees of control over the exploitation of fuels and energy costs. There were thirty-five government agencies with some responsibility for energy policy-making at the national level, according to one analysis by the Office of the Secretary of the Interior.[12]

"Over-all, this is a pluralistic system, characterized by the disjointed, uncoordinated, and piecemeal nature of its decision-making. One consequence is that policy outputs and outcomes from this system often appear to be irrational and/or dysfunctional," observed Professor Irvin L. White, assistant director of the Science and Public Policy Program at the University of Oklahoma.[13]

The "dysfunction" White described was nowhere better illustrated than in the AEC's policy of promoting nuclear reactors to save oil and gas while pushing nuclear explosives to accelerate their recovery.

Insofar as it affected atomic-power development, the Court of Appeals decision had the effect of setting an environmental parameter to the uninhibited development of nuclear-power plants. The decision injected a balance into the agency's regulatory procedures, a consideration which had been lacking—and which reflected a growing concern about a balance between man and nature.

In September, the AEC published an interim statement of its policy and procedures for implementing NEPA in accordance with the court decision. The effect of its new regulations would be to make the agency directly responsible for evaluating the total environmental impact, including the thermal effects, of nuclear-power plants. This impact, the agency added, would be assessed in terms of the available alternatives and the need for electric power.[14]

The AEC, however, was not abandoning its prized role as a champion of atomic-electric power. After acknowledging that "The Commission intends to be responsive to the conservation and environmental concerns of the public," the interim statement asserted: "At the same time the Commission is also examining the steps that can be taken to reconcile a proper regard for the environment with the necessity for meeting the nation's growing requirements for electric power on a timely basis."

This caveat might have given some reassurance to the industry that the power interests would not be forgotten, but there was little doubt that the new regulations would open a Pandora's Box of intervenor issues, ranging from aesthetic and thermal pollution to renewed challenges to the adequacy of the radiation standards. These challenges would inevitably affect the development of the demonstration program which was planned by the AEC, with President Nixon's blessing, to put the breeder reactor on line sometime in 1980.

Sir Gawaine Rides Again

To the dismay of some executives in the utility and supplier industry, Schlesinger began making noises which suggested he would become the Sir Gawaine of the beset environmentalists. He was certainly taking a soft line toward environmentalist intervention, while at the same time forecasting a rosy future for the nuclear electrical industry. The two seemed to be inconsistent to practical power men.

In September 1971 the new chairman flew to Geneva, Switzerland, where the retiring chairman, Seaborg, was presiding at the Fourth United Nations Conference on the Peaceful Uses of Atomic Energy—and where the American Atomic Industrial Establishment was enthusiastically promoting a plutonium-fueled, brave new world, with the cooperation of its counterparts in the Soviet Union and in Western Europe.

Schlesinger met with the agents of the major American utilities and manufacturers of nuclear-reactor hardware at the Intercontinental Hotel and gave them, in his own words, an optimistic assessment of the future of atomic energy. He also held an off-record news conference with several American newsmen, who came away with the impression that he was, indeed, the knight-errant of the environmentalists. It was clear then that he intended to reverse the AEC's spoiler image even if it slowed down the nuclear-reactor program. That was precisely what the industry feared he was going to do. Schlesinger's decision in August not to appeal the Calvert Cliffs decision to the United States Supreme Court had enhanced industry's apprehension about the new, soft line.

The Referee

Perhaps industry's worst fears were confirmed at a banquet of the Atomic Industrial Forum-American Nuclear Society meeting at Bal Harbour, Forida, on October 20, 1971, where Schlesinger publicly unveiled the new approach of the AEC. In its report of the event *Science* magazine said:

> With patrician *froideur,* Schlesinger informed a mass gathering of the nuclear-power industry . . . that from henceforth the AEC would act as the referee of nuclear power, not its promoter. . . . Schlesinger served notice on the nuclear banqueteers that their cozy relationship with the AEC was at an end: The industry should not expect the AEC to fight its battles; it should take its own case to the public as the Sierra Club does.[15]

The essential message Schlesinger broadcast over the public-address system was that the industry could no longer rely on government support in dealing with the environmentalists. It was the AEC's responsibility to develop new technical options and bring them to the point of commercial application, he said; it was not the agency's burden to solve industry's problems in exploiting the applications.

In its role of referee, rather than as partner, the AEC might gain some of the public confidence it had lost over the years of supporting the industry against the people. But that depended on how the agency called the close ones.

"I might add that it is to industry's long-run advantage that the public has high confidence that the AEC will appropriately perform its role in this regard," Schlesinger said.

Since he had taken the AEC chair, Schlesinger said, he had been impressed by the failure "in house" and in the industry to distinguish between the role of industry and the role of the agency. The industry should remember that the AEC does not sell power reactors; that it does not have a responsibility to supply power; and that:

> Finally, and let me underscore this, it is not the AEC responsibility to ignore in your behalf an indication of Congressional intent or to ignore the courts. We have had a fair amount of advice on how to evade the clear mandate of the federal courts. It is advice that we did not think proper to accept.

Beyond this, he said, environmentalists had raised questions that transcended those involved in individual plants—for example, whether society, for environmental reasons, ought to curb its appetite for energy and electricity was "a legitimate social question." The remark was an unheard-of concession from the Establishment point of view, which had tended toward the conviction that only crackpots raised such issues. The new chairman went on:

> It is not unreasonable to question whether neon signs or even air conditioning are essential ingredients in the American way of life. More fundamentally, it is not unthinkable to inquire whether energy production should be determined solely in response to market demand.

That, too, was a revolutionary statement. So far as energy policy was concerned, the AEC would be officially neutral, Schlesinger said.

The Bal Harbour speech expressed a sharp turnabout in the attitude of the Executive Branch toward the role of the AEC. Some of the Commissioners were slow to get the message, but ultimately they would all have to conform to the new policy, which tended to dissolve the government-industry partnership. The role of industry in guiding the attitudes and policies of the Atomic Industrial Establishment was to be sharply curtailed. The role of AEC and, indirectly, of the Joint Committee, in coaching the industry in its relationship with the public was to end. In brief, the industry had lost a coach and gained an umpire. At least, that was what the New Look was supposed to accomplish.

But this still was not the inauguration of an energy policy; rather, it was simply the removal of a barrier to the evolution of one. In the shaping of a national energy policy, there was now an opportunity for the people to be heard.

Backlash

It was not long, however, before the Atomic Industrial Establishment maneuvered to outflank the Calvert Cliffs decision. No sooner had the AEC promulgated revised licensing regulations in September 1971 in conformity with the decision than the industry began to complain about protracted delays in bringing new reactors on line, warning that the inevitable consequence of the new requirements would be brownouts and blackouts. The industry was trying to create an impression that although nuclear power was providing hardly more than 1 per cent of the total electric capacity in the United States, a delay in issuing operating licenses to some new reactors would plunge parts of the nation into darkness. Like magic, the AEC produced figures in support of this warning.

Representative Hosmer of the Joint Committee told the House of Representatives that thirteen completed nuclear-power stations were being kept idle by delays resulting from the new NEPA requirements.[16] Exhibit "A" was the Quad Cities Nuclear Power Station at Cordova, Illinois, on the Mississippi River, near the four cities of Rock Island, Moline, and East Moline, Illinois, and Davenport, Iowa. The United States District Court in the District of Columbia had enjoined the AEC preliminarily from issuing an interim license that would allow the plant's two boiling-water reactor units of 809 megawatts each to run at half-power capacity for testing and power production. This ruling was based on the complaints of the Izaak Walton League of America and the United Auto, Aerospace, and Agricultural Implement Workers of America in one suit and of the State of Illinois in another.

The suits spotlighted a loophole in the AEC's revised licensing regulations: a provision in an appendix that allowed the agency to continue its practice of issuing *interim* operating licenses to nuclear-power plants without either the preparation of a NEPA environmental-impact statement or an opportunity for the public to be heard.

The AEC had announced on March 16, 1971, that it would issue an interim license permitting Quad Cities to operate at half power. But it was not until six weeks later that the joint owners, Commonwealth Edison Company and Iowa-Illinois Gas and Electric Company, revealed their plans to use cooling systems designed to return hot water from the condenser cooling system to the Mississippi, thus threatening to degrade biologically America's greatest river. The disclosure came after the expiration of the thirty-day period in which an intervenor could invoke the jurisdiction of the United States Court of Appeals to halt the projected thermal pollution of the river.

On November 4, 1971, the plaintiffs took their case to the United States District Court, where they contended that the post-Calvert Cliffs licensing regulations under which the AEC proposed to issue the interim license to Quad Cities were invalid on the grounds that the procedures were promulgated without either an environmental statement or a public hearing. Secondly, they asked the court to enjoin the AEC from issuing the interim license to the Quad Cities plant until the required NEPA environmental assessment had been prepared.

Commenting on the first count of the complaint, Judge Barrington J. Parker said the District Court was without jurisdiction to rule on the validity of the AEC's revised licensing regulations, since this was a matter for the Court of Appeals. But commenting on the second count he said:

> This Court views with concern the fact that the manner in which the thermal effluents would be discharged into the Mississippi River (and the alleged dangers of such discharges) was not disclosed until April 30, 1971. Then, the thirty-day period within which the plaintiffs could have become parties and therefore could invoke the jurisdiction of the Court of Appeals had expired.

Thus, Judge Parker ruled on the second count of the complaint that the plaintiffs were entitled to a preliminary injunction restraining the AEC from issuing the interim operating license. The AEC immediately appealed the ruling and for a while it looked like a long struggle.

The power industry intensified the drum beat of warnings of a power shortage, unless something was done. In the Quad Cities case, Iowa-Illinois Gas and Electric faced the summer of 1972 with a total generating capacity 45 megawatts less than the predicted peak load if its share of power from Quad Cities Unit 1 was not available. It was fortuitous, indeed, that this plant, begun in 1966, was being completed just in the nick of time. Commonwealth Edison would have a reserve margin of only 5.4 per cent above summer peak demand without Quad Cities, according to the AEC estimate.

Members of the Nixon Administration, the Joint Committee, and other officials moved to rescue the beleaguered nuclear-power plants from the tentacles of NEPA. Representative John D. Dingell of Michigan drafted an amendment to NEPA allowing the AEC to issue interim operating licenses until the end of October 1973 in certain instances where a power emergency appeared imminent. Representative Hosmer described the Dingell bill as "a kind of suspension, for a particular purpose, of the Environmental Protection Act under very, very limited and discreet conditions." [17] Hosmer, himself, was busy drafting legislation that would also accomplish this result.

On another front, a rescue mission was launched by Senator Howard A. Baker, Jr., of Tennessee, a member of the Joint Committee and also of Senator Muskie's Subcommittee on Air and Water Pollution. He added an amendment to Muskie's Water Quality Bill (with Muskie's consent) exempting the AEC from complying with NEPA's Section 102—the section requiring federal agencies to assess the impact of their actions or projects affecting the environment. The amendment would return jurisdiction of thermal pollution of waterways by nuclear-power plants to the states—where the AEC contended it belonged.

Both Dingell and Baker considered themselves concerned environmentalists, but when it came to a showdown between the demand for more power and the prevention of thermal pollution of rivers and lakes, they appeared to be on the side of more power. Both legislators were members of the United States delegation to the United Nations Conference on the Human Environment, held at Stockholm, Sweden, in June 1972.

A third bill sponsored by Hosmer (H. R. 14655) with the blessing of the Joint Committee and the White House amended the Atomic Energy Act of 1954 to authorize the AEC to issue interim operating licenses in special cases during the expected power shortage period ending October 30, 1973. This amendment seemed to be less of a threat to NEPA than the Dingell or Baker amendments, but some environmental experts feared it would have the same effect. The environmental groups' victory in the Quad Cities case notwithstanding, passage by Congress of the Hosmer amendment, which was signed into law by President Nixon in June, marked a turning away from the Calvert Cliffs decision and the resumption of the old, laissez-faire policy toward the thermal-pollution effects of nuclear-power plants. Slowly but surely the AEC was freeing itself from the strictures of NEPA.

Meanwhile, the Quad Cities case was settled. The utilities agreed to construct a $30-million, serpentine canal into which hot water from the condensers would be sprayed, thus releasing its heat to the atmosphere. Until this closed-cycle system was finished, however, the utilities could discharge hot water into the river through a diffuser pipe on the bottom. The Izaak Walton League, the UAW, and the State of Illinois then withdrew their suits for a permanent injunction, and the AEC dropped its appeal from the preliminary one.

With Congressional amendment of the Atomic Energy Act, the power-shortage issue seemed to vanish, and NEPA had been compromised—at least in its application to nuclear power. The balance between nuclear-power production and the protection of the environment remained both unresolved and uncertain. Achieving the balance was not so much a question of scientific, legal, or moral judgment as it was a struggle between environmental and power interests.

On the theory of a power shortage, the AEC, the Joint Committee, and the White House had promoted quick, interim licensing of new plants, overriding environmental and safety concerns. Still being debated at AEC hearings at the time was the vital question of whether an emergency core-cooling system could be devised to protect the public against the disastrous consequences of a core meltdown, in the event of a loss of coolant.

It was clear that if the environment were to be protected, citizens would have to do it; that if public health and safety were to be guaranteed, citizens would have to see to it. The agency to which these responsibilities had been entrusted was busy serving the interests which it was supposed to be regulating.

At the end of June 1972, the United States Court of Appeals at Chicago enjoined the AEC from issuing a temporary operating license to Point Beach No. 2 nuclear-power unit of the Wisconsin Electric Power and Wisconsin-Michigan Power Companies near Two Rivers, Wisconsin. The license would have allowed the plant a daily discharge of a half-billion gallons of hot water into Lake Michigan. The AEC sought to issue the license while environmental groups were appealing the thermal-pollution issue to the AEC's Safety and Licensing Appeals Board.

In rushing to the relief of the utilities, the AEC had quickly slipped back into its old role of promoter first, regulator of atomic energy second. But concerned citizens could not be defeated.

The nuclear-power rebellion rolled on. It was the only real challenge to a reckless nuclear-energy policy that the Atomic Industrial Establishment had to face.

Reference Notes

Chapter 1. Another Kind of Fire

1. "Civilian Nuclear Power," a Report to the President. Atomic Energy Commission, 1962.
2. *Journal of the Sanitary Engineering Division,* Proceedings of the American Society of Civil Engineers, June 1969. Authors: J. G. Terrill, Jr., C. L. Weaver, E. D. Harward, and J. M. Smith.
3. *Nuclear Safety,* Vol. 12, No. 5, September–October 1971.
4. *Hearings,* Joint Committee on Atomic Energy, April 24, 1969. "Selected Materials on Environmental Effects of Producing Electric Power."
5. *Hearings,* Senate Committee to Investigate Construction of Atomic Generating Plants in Pennsylvania, August 20, 1970.

Chapter 2. A Pattern for Plunder

1. Report of the Joint Committee on Atomic Energy, "Nuclear Power Economics, 1962–67," February 1968 (p. 52).
2. *Hearings,* JCAE, Participation by Small Electrical Utilities in Nuclear Power, 90th Congress, 2nd Session, Part 1. "Separate Views of Representatives Holifield and Price on HR 9757," Appendix 11 (April 30–May 3, 1968).
3. *Ibid.*
4. United States Court of Appeals, District of Columbia Circuit, Nos. 21,706 and 21,844, Cities of Statesville, etc., and Power Planning Committee, etc.
5. United States Court of Appeals, Nos. 21,707 and 21,882, Municipal Electric Association and Eastern Maine Electric Cooperative, etc.
6. *Hearings,* JCAE, 90th Congress, 2nd Session, Part 2, June 13, 1968.
7. *Hearings,* JCAE, Prelicensing Anti-trust Review of Nuclear Power Plants, 91st Congress, 1st Session, Part 1 (p. 176), Appendix 1 (November 18–20, 1969).
8. *Hearings,* JCAE, 90th Congress, 2nd Session, Part 2, June 13, 1968.
9. *Ibid.*
10. *Hearings,* JCAE, 90th Congress, 2nd Session, Part 1, April 30, May 1, 2, and 3, 1968.
11. *Ibid.* (p. 23).
12. JCAE Report, "Nuclear Power Economics" (p. 10).
13. *Ibid.* (p. 33).
14. Letter to President Nixon from Alex Radin, General Manager, American Public Power Association.

300 / Reference Notes

15. "Competition in Energy Markets," an analysis by National Economic Research Associates, Inc., prepared for the Senate Subcommittee on Antitrust and Monopoly, Committee of the Judiciary, May 1970.
16. *Ibid.*
17. *Ibid.*
18. "Effects of Thermal Discharges on the Mass-Energy Balance of Lake Michigan," J. G. Asbury, Center for Environmental Studies, Argonne National Laboratory, 1970.

Chapter 3. The Radiation Heresy

1. International Commission on Radiological Protection Publication No. 6. Report, Pergamon Press. 1964.
2. Symposium on Nuclear Power and the Public, University of Minnesota, October 11, 1969.
3. Report of Warrent Una, *The Washington Post,* as published in the *Bulletin of the Atomic Scientists,* November 1967.
4. "Evidence for Low-Level Radiation Effects on the Human Embryo and Fetus," presented at the Hanford Symposium on the Radiation Biology of the Fetal and Juvenile Mammal, May 1969.
5. B. MacMahon, *Journal of the National Cancer Institute,* 28, 1773, 1962.
6. *Power Generation and Environmental Changes,* edited by David A. Berkowitz and Arthur M. Squires, Cambridge, Mass.: MIT Press, 1971. "Investigation of the Effects of X-Ray Exposure of Human Female Fetuses as Measured by Later Reproductive Performance. Interim Summary." Mary B. Meyer, Earl L. Diamond, and Timothy Merz, School of Hygiene and Public Health, The Johns Hopkins University.
7. E. J. Sternglass, *Low Level Radiation.* New York: Ballantine Books. 1972.
8. *Bulletin of the Atomic Scientists,* September 1970.
9. Congressional Record, July 29, 1969, H 6505.
10. *American Journal of Public Health,* February 1970.
11. *Science,* October 10, 1969 (p. 196).
12. *Ibid.*
13. Personal communication to Sternglass.
14. *Bulletin of the Atomic Scientists,* June 1969 (p. 27).
15. *Ibid.*
16. *Radioactive Waste Discharges to the Environment from Nuclear Power Facilities.* U.S. Department of Health, Education, and Welfare, March 1970.
17. Norman C. Dyer and A. Bertrand Brill, "Fetal Radiation Dose from Maternally Administered Fe 59 and I 131." Radiation Biology of the Fetal and Juvenile Mammal, Hanford Symposium, May 1969.
18. Alice Stewart and G. W. Kneale, *Lancet,* June 6, 1970.
19. Sixth Annual Health Physics Society Topical Symposium, November 2–5, 1971.
20. Presented at Conference on Planning and Epidemiological Study of Pollution Effects, Sixth Berkeley Symposium on Mathematical Statistics and Probability, Berkeley, Calif., July 19–22, 1971.

Chapter 4. The Ghost of Galileo

1. *Bulletin of the Atomic Scientists,* September 1970.
2. *Ibid.*
3. *The New York Times,* January 26, 1971.
4. *Hearings,* JCAE, Environmental Effects of Producing Electric Power, 91st Congress, 2nd Session, Part 2, Vol. II, January 27, 1970.
5. *Ibid.*
6. "Studies of Radium-Exposed Humans, II," John W. Gofman and Arthur R. Tamplin, submitted to JCAE, January 28, 1970.
7. *Ibid.*
8. Statement by Edward Teller at New York–New Jersey Air Pollution Conference, January 1967.
9. "What Shall Be Done About the Maximum Permissible Dose?" paper by Teller submitted to Senator Mike Gravel, December 19, 1969.
10. Teller's response to questions by Senator Gravel accompanying correspondence of December 19, 1969.
11. *Hearings,* JCAE, Environmental Effects of Producing Electric Power, 91st Congress, 1st Session, Part 1, Vol. I, October 30, 1969.
12. *Ibid.*
13. Staff Report on Allegations made by Drs. Tamplin and Gofman of Censorship and Reprisal by the AEC and the Lawrence Radiation Laboratory at Livermore, July 21, 1970.
14. *Science,* August 28, 1970.
15. Letter from Ralph Nader to Senator Edmund S. Muskie, July 5, 1970.
16. *Hearings,* JCAE, Environmental Effects, etc., 91st Congress, 1st Session, Part 1, Vol. I.

Chapter 5. The Intervenors

1. "National Radiation Health Standards: A Study in Scientific Decision Making." *Atomic Energy Law Journal,* Vol. 6, No. 3, fall, 1964.
2. *Ibid.*
3. Statement of Ernest C. Tsivoglou to the Minnesota Pollution Control Agency.
4. Annual Report to Congress, AEC, 1970 (p. 264).
5. "Radioactive Pollution Control in Minnesota," Tsivoglou report to Minnesota Pollution Control Agency.
6. Statement of Intervenors Kenneth Dzugan, George B. Burnett II, and Theodore I. Pepin in the Monticello Nuclear Power Station Operating License Hearing, August 24, 1970.
7. Symposium on "Nuclear Power and the Public," University of Minnesota, October 10–11, 1969.
8. Appendix A, "Dunster-Tsivoglou correspondence, 1969," and Craig Hosmer, "Federal Pre-emption of Regulation of Release of Radionuclides from Nuclear Power Plants: A Legal History and Rationale," in *Nuclear Power and the Public,* edited by Harry Foreman, M.D. University of Minnesota Press, June 1970.
9. *Ibid.*
10. Letter of April 2, 1971, from Robert H. Engels to Dr. Howard A. An-

derson, Chairman, Minnesota Pollution Control Agency.
11. *The New York Times,* June 7, 1971.
12. *Ibid.*
13. *Hearings* before the Joint Committee on Atomic Energy, June 22, 1971.
14. *Ibid.*

Chapter 6. Boot Hill

1. "Report on the Examination of the Waste Treatment and Disposal Operations of the National Reactor Testing Station, Idaho Falls, Idaho." April 1970. Federal Water Quality Administration submission to Senator Frank Church of Idaho.
2. AEC draft, Environmental Impact Statement on the National Radioactive Waste Repository at Lyons, Kansas.
3. *Ibid.*
4. *Ibid.*
5. *Authorization Hearings,* JCAE, 92nd Congress, 1st Session, March 1971.
6. Letter dated February 3, 1971, from Hollis M. Dole, Assistant Secretary, Department of the Interior, to John A. Erlewine, Assistant General Manager for Operations, AEC.
7. Letter dated April 29, 1971, from Arthur H. Wolff, Acting Deputy Commissioner, Radiation Office, EPA, to John A. Erlewine, Assistant General Manager for Operations, AEC. (*Congressional Record,* May 6, 1971, S 6321.)
8. Comment by Floyd L. Culler, *Bulletin of the Atomic Scientists,* June 1971.
9. *Ibid.*
10. Congressional Record, July 15, 1971.
11. *Ibid.*

Chapter 7. Plowshare

1. E. A. Martell, "Plowing a Nuclear Furrow." *Environment,* April 1969.
2. Atlantic and Pacific Interoceanic Canal Study Commission, *Final Report,* December 1930.
3. *Ibid.*
4. Peaceful Applications of Nuclear Explosives—Plowshare. *Hearing,* Joint Committee on Atomic Energy, 89th Congress, 1st Session, January 5, 1965.
5. *Ibid.*
6. *Ibid.*
7. Paper by R. S. Davidson, Battelle Memorial Institute, Columbus Laboratories, Smithsonian Institution Panama Conference on Tropical Biology, Panama City, R.P.
8. *Ibid.*
9. "Prediction of External Gamma Dose from Nuclear Excavation of a Sea Level Canal." *BioScience,* March 1969.
10. E. A. Martell, "Plowing a Nuclear Furrow." *Environment,* April 1969.
11. Atlantic and Pacific Canal Study Commission, *Final Report.*

12. *Ibid.*
13. Glenn T. Seaborg letter to Chairman Anderson, July 7, 1970.

Chapter 8. Bonanza from Bombs

1. M. King Hubbert, "Resources and Man." National Academy of Science-National Research Council. Symposium, 1969.
2. In address, "The Energy-Environment Equation," Fourth United Nations Conference on the Peaceful Uses of Atomic Energy, Geneva, Switzerland, September 1971.
3. Statement of Executive Committee, Pittsburgh Chapter, Federation of American Scientists, at Lock Haven State College Forum, February 16, 1968.
4. Associated Press report by Ralph Dighton on AEC Study by Dr. Norman French, ecologist, University of California at Los Angeles.
5. J. Wade Watkins and C. C. Anderson, "Potential of Nuclear Explosives for Producing Hydrocarbons from Deposits of Oil, Natural Gas, Oil Shale and Tar Sands in the United States." United States Bureau of Mines.
6. *Ibid.*
7. *Ibid.*
8. *Ibid.*
9. *Ibid.*
10. AEC paper presented to Pacific Northwest Region Conference, American Institute of Mining, Metallurgical, and Petroleum Engineers, Seattle, Washington, April 21–22, 1966.
11. "An Analysis of Nuclear Gas Stimulation and the Program Required for Its Development," G. C. Werth *et al.,* AEC.
12. Statement of John M. Nassikas, Chairman, Federal Power Commission, JCAE *Hearing,* March 23, 1971.
13. AEC Annual Report to Congress, 1967.
14. *Hearings,* Underground Uses of Nuclear Energy, Vol. I, Subcommittee on Air and Water Pollution, Senate Committee on Public Works, 91st Congress.
15. Paper presented to Society of Petroleum Engineers, Houston, Texas, September 30, 1968, by Charles F. Smith and Floyd E. Momyer, Lawrence Radiation Laboratory.
16. "Preliminary Assessment of the Radiological Implications of Commercial Utilization of Natural Gas from a Nuclear-Stimulated Well." D. G. Jacobs and E. G. Struxness, Oak Ridge National Laboratory and C. R. Bowman, El Paso Natural Gas Company.
17. *Ibid.*
18. Nuclear Explosion Services for Industrial Applications, *Hearings* before the JCAE, 91st Congress, 1st Session, May 8, 9, and July 17, 1969.
19. *The New York Times,* April 6, 1969.
20. Gas Utility and Pipeline Projections, 1966–1975. American Gas Association, October 1966.
21. *Ibid.*
22. *Ibid.*
23. Project Ketch Feasibility Study, AEC, U.S. Bureau of Mines, Columbia Gas System Service Corporation, 1967.

304 / Reference Notes

24. Annual Report to Congress, AEC, 1970.
25. *Authorization Hearings,* JCAE, 1971 Fiscal Year.
26. Structural Response Studies, Project Rulison, Nevada Operations Office, AEC.
27. Associated Press report, Mercury, Nevada, December 19, 1970.
28. *Authorization Hearings,* JCAE, 1971 Fiscal Year.
29. *Ibid.* (p. 625).
30. *Authorization Hearings,* JCAE, 1972 Fiscal Year.
31. *Ibid.*

Chapter 9. The Battle of the Breeder

1. Interview by Novosti Press Agency, *Bulletin of the Atomic Scientists,* October 1971.
2. *Sourcebook of Atomic Energy,* Samuel Glasstone, USAEC, 3rd ed., 1967 (p. 540).
3. *Bulletin of the Atomic Scientists,* October 1971.
4. Statement of the Federal Power Commission Chairman, March 19, 1971, to the JCAE.
5. *Bulletin of the Atomic Scientists,* October 1971.
6. *Ibid.*
7. Statement of the Federal Power Commission Chairman, March 19, 1971, to the JCAE.
8. *Bulletin of the Atomic Scientists,* October 1971.
9. Allen L. Hammond, "Breeder Reactors: Power for the Future." *Science,* Vol. 174, November 19, 1971.
10. Fourth International Conference on Plutonium and other Actinides. October 5, 1970.
11. *Bulletin of the Atomic Scientists,* September 1971.
12. Ann Mozley, "Change in the Argonne National Laboratory, a Case Study." *Science,* Vol. 173, October 1, 1971.
13. Fourth International Conference on Plutonium and Other Actinides. October 5, 1970.
14. Address to the National Academy of Sciences Symposium, "Energy for the Future," April 1971.
15. "Potential Nuclear Power Growth Patterns," AEC, WASH 1098, December 1970.

Chapter 10. Milestone at Calvert Cliffs

1. Staff Report, Administrative Conference of the United States: "Licensing of Nuclear Power Plants by the AEC, April 1967."
2. *Ibid.*
3. *Hearings,* JCAE, AEC Licensing Procedure and Related Legislation, Subcommittee on Legislation, 92nd Congress, 1st Session, Part 2, June 22, 23; July 13, 14, 1971.
4. Letter dated June 8, 1971, to Torbert H. Macdonald, Chairman, Subcommittee on Communications and Power, Committee on Interstate and Foreign Commerce, House of Representatives.
5. S. David Freeman, "Toward a Policy of Energy Conservation." *Bulletin of the Atomic Scientists,* October 1971.

6. From a speech on May 11, 1971, to the New York Metropolitan Section of the American Nuclear Society.
7. *Hearings,* JCAE, AEC Licensing Procedure and Related Legislation, Subcommittee on Legislation, 92nd Congress, 1st Session, Part 1, June 22, 1971.
8. *Hearings,* JCAE, Environmental Effects of Producing Electric Power, 91st Congress, 2nd Session, Vol. 1, Part 2 (p. 1333).
9. *Hearings,* JCAE, Environmental Effects of Producing Electric Power, 91st Congress, 2nd Session, Vol. II, Part 2 "Thermal Effects Studies by Nuclear Power Plant Licensees and Applicants, 1969," an AEC Survey.
10. Federal Power Commission Staff Study, Disposal of Waste Heat, 1969.
11. *Bulletin of the Atomic Scientists,* October 1971.
12. *Ibid.*
13. Irvin L. White, "Energy Policy Making." *Bulletin of the Atomic Scientists,* October 1971.
14. Federal Register, Vol. 26, No. 175, September 9, 1971.
15. *Science,* October 29, 1971.
16. Congressional Record, May 3, 1972, H4034.
17. *Ibid.*

Index

Abrahamson, Dean E., 255-57
Adams, Franklin S., 216-17
Advisory Committee on Reactor Safeguards (ACRS), 263-64
Aiken, George D., 21, 29-30, 35, 36-37, 38
Aiken-Kennedy bill, 35, 37, 38
"Alpha" wastes, 150
American Association for the Advancement of Science (AAAS), 98-99
American Power Conference (1971), 243
American Public Power Association, 39, 42
Anderson, C. C., 202
Anderson, Wendell R., 133
Angino, Ernest, 165
Antiballistic Missile System (ABM), 68
Argonne National Laboratory, 46, 57, 248
Arias, Arnulfo, 196
Aristarchus, 90
Arraj, Alfred A., 224
Artsimovich, Lev A., 232, 238
Atkinson, C. H., 204
Atlantic and Pacific Interoceanic Canal Study Commission, 179-180, 185-86, 188-91, 193; Technical Associates for Geology, Slope Stability, and Foundation, 196-97
Atmospheric testing, 63-66, 188
Atomic Bomb Casualty Commission (ABCC), 87

Atomic energy, peaceful uses of. *See* Project Plowshare
Atomic Energy Act (1954), 26, 28-29, 35, 39-40, 55, 123, 140, 224, 261, 263, 295-96
Atomic Energy Commission (AEC), ACBM, 97; governing regulations, 278; image, 109; and inspection, 130; leadership change, 286-87; licensing policy, 29-32, 36, 39, 131, 132, 135, 263-65; and LMFBR, 250-57; and NEPA, 276-78; and oil and gas resources, 199, 202, 227-30; and radioactive wastes, 149-52, 155, 157-71; radiation standards, 11, 21-23, 48-56, 84-86, 111-12, 122-23, 125-33, 136-47; role of, 114; "Truth Squad," 71-72; *see also* Project Plowshare
Atomic Industrial Establishment, 10, 13, 27
Atomic Safety and Licensing Board, 115, 116, 136
Austral Oil Company, 220, 222
Ayers, Thomas G., 267

Bailey, Daniel A., 215
Baker, Howard A., Jr., 295, 296
Balboa, Vasco, 182
Baltimore Gas and Electric Company (BG&E), 269, 272, 276-78, 287
Banebury Test, 226-27
Barger, Mrs. Walter, 211
Battelle Memorial Institute, 187

308 / Index

Batzel, Roger E., 99, 100-101
Bazelon, David L., 32-33
Benedict, Manson, 249-50
Bingham, Jonathan B., 68-69
Biskis, Birute O., 57
Blanchard, Harry, 268
Bond, Victor, 224
Bossert, Max, 215
Brave New World (Huxley), 216
Breeder Reactor Corporation, 258
Breeder reactors, 4-5, 19, 20, 45, 232-33, 237, 242-43, 250-52, 258. *See also* Liquid Metal Fast Breeder Reactor
Brodsky, Allen, 213
Brookhaven National Laboratory, 18, 118-21
Brookings Institution, 41
Brown, Harold, 174, 175
Brown, Howard C., Jr., 105-107
Buggy I test, 188
Bunau-Varilla, Philippe, 193
Burnett, George B., II, 130
Burr, Spencer, 10

Cabriolet test, 188
Calvert Cliffs case, 269-71, 274-87
Calvert Cliffs Coordinating Committee (Md.), 277
Cancer, and radiation, 57-59, 61-63, 85-91, 116, 117
Cape Karauden project (Australia), 176-77
Capucci, Victor, Jr., 10
Carl, Mrs., 267-68
Central America, and Plowshare, 177-98
Chandler, W. W. Jr., 154
Chase, Helen C., 64
Cherry, Myron M., 138-39
Cherry, William R., 240
Chesapeake Environmental Protection Association (CEPA), 270-273, 281
Choco Indians (Central America), 190, 191
Church, Frank, 161
Citizens Committee for Environmental Concern (Pa.), 7-9, 12-13
"Clementine" reactor, 243
Closed-cycle cooling, 267
Coal, 26, 42, 199

Coleridge, Samuel Taylor, 228
Collins, Michael F., 33-34, 36, 37
Colombia, 184-85, 189-91, 192, 194
Columbia Gas System Service Corporation, 208-12, 214-15, 220
Columbus, Christopher, 90
Combustion Engineering Company, 270
Committee on Geologic Aspects of Radioactive Waste Disposal, 159
Committee on Radioactive Waste Management, 161
Commoner, Barry, 250
Commonwealth Edison Company (Ill.), 16, 258, 293, 295
Connecticut Yankee Atomic Power Plant, 37
Consumers Power Company, 135, 138-40, 142, 275, 281-82
Copernicus, Nicolaus, 90, 91
Council of Economic Advisers, 42
Crick, Francis H. C., 116
Culler, Floyd L., 167
Cuna Indians (Panama), 181-82, 183, 191

Daniels, Joan, 7, 9, 268
Danny Boy test, 186
Deale, Valentine B., 129
DeGroot, Morris H., 77-78
De revolutionibus orbium coelestium (Copernicus), 91
Detroit Edison, 143
Deuterium, 234
Devitt, Edward J., 126
Dieckamp, Herman, 21-22
Dingell, John D., 295, 296
DiStadio, Gus, 10, 11
DNA, 61
Docking, Robert, 156, 158-59, 162, 163, 166, 168-69
Dole, Hollis M., 166
Dole, Robert, 169
Doub, William O., 286-87
Downs, Hugh, 72
Dresden I plant (Ill.), 22, 73-77, 128
Dresden II plant (Ill.), 16
Dressler, Mrs. Roy, 157-58
Dressler, Roy, 157, 158, 165
Druckman, Aaron, 217
Dubos, René, 71
DuBridge, Lee A., 36

Duke Power Company case, 31
Dumont, Martin G., 221
Dunster, H. J., 132-33
Dyson, Freeman J., 70-71
Dzugan, Kenneth, 130

Eastern Maine Electric Cooperative, 34
EBR-I reactor, 244
EBR-II reactor, 244
Edison Electric Institute, 37, 38
Einstein, Albert, 90
Eisenhower, Dwight D., 51, 115, 286
El Paso Natural Gas Company, 203-205, 207
Engels, Robert H., 133, 134
English, Spofford, 82
Environmental Protection Agency (EPA), 52, 115; Radiation Office, 166
Environmental Science Services Administration, 190
Erlewine, John A., 169
Evans, Robley D., 95-96

Fallout. *See* Radiation
Federal Power Commission (FPC), 44, 204, 238, 273-74, 276
Federal Radiation Council (FRC), 51, 93, 115, 145, 147; Radiation Protection Guides, 84
Federal Water Pollution Control Act, 141
Federal Water Quality Administration (FWQA), 149-50
Federation of American Scientists, 213-14
Ferber, Gilbert J., 188
Fermi I reactor, 4, 15, 164, 244
Fermi II reactor, 143
Fermi, Enrico, 4
Finkel, Miriam P., 57
FLASK test, 197
Flood, Dan, 24
Ford, Norman C., 239
Fort Peck experiment, 184
Fossil-fuel industry, 42-43, 44
Fourth International Conference on Plasma Physics and Nuclear Fusion Research, 236-37
Fourth International United Nations Conference on the Peaceful Uses of Atomic Energy, 236, 290
Fowler, Kenneth, 237
Freeman, S. David, 288
Fulmer, Eugene M., 215, 219
Fusion reactors, 233-34, 237-38

Galen, 90
Galilei, Galileo, 49, 90
Garner, Marie, 215
Gasbuggy experiment, 204-207, 225
Geesaman, Donald P., 100, 103, 247-48
General Electric, 40-41
Gerdes, Robert H., 37
Gerusky, Thomas M., 8
Goethals, George W., 178
Gofman, John W., 17-18, 19, 21, 22, 23, 48-49, 50, 83-93, 96-101, 103-104, 144
Gould, Roy W., 237
Gravel, Mike, 96
Great Britain, 194, 233, 235, 242, 275
Great Lakes, 156
Greek fire, 228-29

Halepaska, John, 158
Hambleton, William W., 155, 156, 158, 159, 162-63, 166-67
Hamburger, Richard, 207
Hamilton, Alexander, 217
Hammond, Allen L., 246
Harvey, Arthur, 154
Harvey, William, 90
Hay, John, 193, 194
Hay-Bunau-Varilla Treaty (1903), 179, 194
Hay-Pauncefote treaties (1901-1902), 194
Hayes, Harold, 68
Heisler, Francis, 10
Heron, 174
Hiroshima, 57, 72, 87, 92, 121
Hobbs, Davis R., 10
Holifield, Chet, 28, 39, 68, 101, 260-262, 264, 265, 269, 271, 287
Hosmer, Craig, 39, 55, 107, 132-134, 167, 168, 231, 232, 260-61, 263, 293, 295-96
Hull, Andrew P., 77

310 / Index

Humble Oil Company, 43
Hurricane Betsy, 18
Hutchinson, George, 68

Infant mortality rates, and radiation, 59-60, 63-65, 69-80, 118-19
Ingles, David R., 191-92
Iobst, Fred, 210-14
Iowa-Illinois Gas & Electric Company, 293, 294
International Commission on Radiological Protection (ICRP), 51, 52-53, 54, 85, 91, 93, 112, 132, 133, 145, 147, 259

Jablon, 121
Japan, 177; *see also* Hiroshima, Nagasaki
Johnson, Albert W., 219
Johnson, Gerald, 180
Johnson, Lyndon B., 179, 185
Joint Committee on Atomic Energy, 3, 14, 21, 26, 27, 28-29, 38, 39, 147, 153, 166-68, 230, 287-88
Jones, Clifford, 218

Kalstein, Marvin, 121-22
Kane, Joseph W., 239
Kansas Geological Survey, 155, 158, 159, 165-67, 170
Kato, 121
Kaye, Martin B., 78
Kelly, John S., 225, 228
Kennedy, Edward M., 35-36
Kennedy, John F., 6, 27, 244, 286
Kennedy, Robert F., 35
Kepler, Johannes, 90
Kerr-McGee Corporation, 43
Killian, James R., Jr., 114
Kirkpatrick-Howat, Y., 272, 273
Kneale, G. W., 62, 76, 90, 121
Kury, Franklin L., 8

Laird, Melvin R., 11
Larson, Clarence E., 45, 199
Laurence, Ernest O., 175
Lave, Lester B., 78
Lawrence Radiation Laboratory, 57, 204, 205-206, 209
Lee, Bryan, Jr., 9-13, 15
Lee, Bryan, Sr., 10

Leinhardt, Samuel, 78
Leisk, C. W., 228
Lesseps, Ferdinand de, 193
Leukemia, and X-rays, 61-64, 87, 117, 121
Like, Irving, 120
Liquid Metal Fast Breeder Reactor (LMFBR), 4, 20, 45, 50, 237, 246, 247, 250-57.
List, Robert J., 188
Lloyd Harbor Study Group (N.Y.), 116-18, 120, 121, 267
Long Island Lighting Company, 115-116, 122

MacMahon, Brian, 61, 62, 63, 87, 92
McMillan, Edwin M., 247
Magnetohydrodynamics (MHD), 241
Maine Yankee Atomic Power Company, 34
Malcolm, Jess W., 274-75
Marshall, Islands, 92
Martell, E. A., 192
Massachusetts Institute of Technology, 190
May, Michael S., 98-99, 104
Mead, Margaret, 250
Meinel, Aden B., 239
Meinel, Marjorie, 239
Merritt, Grant J., 134
Meshoppen reactor, 6-7, 18-21
Michigan, Lake, 140, 142
Miller, Vern, 169
"Miniata" test, 228
Minnesota Environmental Control Citizens Association (MECCA), 128, 133
Minnesota Pollution Control Agency, 55, 133, 134, 144
Mohr, W. D., 139-40
Monticello plant (Minnesota), 128, 131, 133-34
Moorhead, William S., 94
Morgan, Karl Z., 71, 93-95
Muller, Hermann J., 56, 57, 60, 66
Municipal Electric Association of Massachusetts, 31, 33, 37
Muskie, Edmund S., 86, 89, 101, 104, 205, 295

Nader, Ralph, 101
Nagasaki, 72, 87, 92, 121

Nassikas, John S., 239
National Academy of Sciences—National Research Council, 151, 159
National Advisory Committee on Radiation, 114
National Aeronautics and Space Administration, 170
National Council on Radiation Protection and Measurements, 51, 52, 84-85, 92, 93, 112, 145-46, 147
National Economic Research Associates (NERA), 43, 44
National Environmental Policy Act (1969), 136, 141, 230, 250-251, 253, 258, 260, 276, 279-85, 293-96; and AEC, 276-78
Natural gas, 44, 199-201; and nuclear stimulation, 203-31
Natural Resources Defense Council, 250
Nautilus (submarine), 40
Nedzi, Lucien N., 68
Nevada nuclear tests, 186-88
Nevada Test Site, 65, 175, 199-200, 226, 256
Newton, Isaac, 90
Niagara Falls, 156
1984 (Orwell), 216
Nixon, Richard M., 46, 51, 68, 166, 242, 243, 248, 250, 286, 287, 289, 296
Nixon Administration, 115, 170, 227
Northern States Power Company, 55, 122-28, 133
Nuclear fusion research, 234-36
Nuclear Non-Proliferation Treaty (1969), 186, 192
Nuclear Power and the Environment (AEC booklet), 106
Nuclear reactors, 4
Nuclear stimulation, 200-201; and natural gas, 203-31
Nuclear Test Ban Treaty (1963), 9, 57, 61, 64, 186, 187, 192, 197

Oak Ridge National Laboratory, 151-53, 155-58, 160, 162-64, 190
Office of Management and Budget (OMB), 227, 228

Office of Science and Technology, 43, 280, 287
Ohme, William, 10
Oil resources, 199-201
Orlofsky, Sy, 216-17

Packard, Russell, 138
Palanquin nuclear test, 187
Palisades plant (Mich.), 135-42, 275-276, 281
Panama, 177-83, 189-91, 193-97
Panama Canal, 177-78, 193-95
Panatomic Canal study (Central America), 178-86, 196-98
Panel on Disposal in Salt of the National Academy of Sciences Committee on Radioactive Waste Management, 159, 161
Parker, Barrington J., 294
Pasteur, Louis, 90
Pastore, John O., 39, 147, 168, 169, 180, 219
Pauling, Linus, 57, 60, 65-66
Pearson, James B., 169
Peconic River, 120
Pedrarias, 182
Pendleton, Robert C., 71
Pennsylvania State Senate Select Committee (hearing), 12-13, 14-15, 17, 21, 23, 53
People Against Ketch (Pa.), 213-15, 217-19
Pepin, Theodore J., 130
Pile, Walter, 154
Pilgrim Nuclear Power Plant (Boston Edison Company), 36
Pleistocene ice ages, 156
Plutonium, 247-48
Point Beach Nuclear Plant Unit Two (Wisc.), 266, 297
Pollution, thermal, 46, 136-37, 141, 259, 275
Post, Richard F., 236
Price, Harold L., 146
Price, Melvin, 28, 39, 167-68
Price-Anderson Act (1957), 18
Project Carryall (Calif.), 176
Project Chariot (Alaska), 176
Project Ketch (Pa.), 207-20
Project Plowshare, 104, 172-77, 199; in Central America, 177-98;

312 / Index

Project Plowshare (cont'd.)
and Nevada tests, 186-88; see also Nuclear stimulation
Project Rio Blanco (Colo.), 229-30
Project Rulison (Colo.), 220-26, 227
Project Salt Vault, 155, 162

Quad Cities case, 293-96
Quale, John G., 265-66

Radford, Edward P., 270-71
Radioactive wastes, disposal of, 148-171
Radiation, and cancer, 57-59, 61-63, 85-91, 116-17; and infant mortality, 59-60, 63-65, 69-80, 118-19; see also X-rays
Rainier test, 175
Ramey, James T., 38, 96-97, 105, 107, 264-65
Reed, George F., 18
Refuse Act of 1899, 251, 285
Reorganization Plan No. 3 (1970), 51
Rex, Robert W., 240-41
Reynolds, Martha, 136
Rickover, Hyman G., 40
Rinehimer, Angela, 10
Robespierre, Maximilien de, 143
Robles, Marco Aurelio, 195-96
Rogers, Lester, 146
Roy, William R., 168
Russia. See Soviet Union

Sagan, Leonard, 71-72
Sailor, Vance L., 119-21
Sandler, Carl L., 192
Saylor, John P., 214-15
Schlesinger, James R., 170, 286, 290-92
Schooner test, 188
Scientists' Institute for Public Information (SIPI), 250-53, 255
Seaborg, Glenn T., 6, 97-98, 105, 136, 137, 161, 189, 197-98, 247, 249, 252, 264, 286, 290
Second International Conference on the Peaceful Uses of Atomic Energy, 175
Securities and Exchange Commission (SEC), 30, 32-34
Sedan test, 186, 188, 197

Shafer, Raymond P., 209, 212, 217, 219, 220
Shapp, Milton J., 219
Shaw, Milton, 21, 38, 127, 248, 249
Shelburne, Tom, 10
Sheret, Virginia, 10
Sherwood, Arthur W., 275
Shippingport reactor (Pa.), 76
Shore, Ferdinand J., 77
Shriver, Garner E., 169
Simonton, Fred G., 34-35
Simpson, John A., 136, 137
Simpson, John W., 243
Sinderman, Roger W., 138-39
Skubitz, Joe, 166, 168, 169
Smith, Sam, 204
Solar energy, 238-40
Soviet Union, 63, 64, 66, 232, 233, 235, 236, 242, 290
Spain, 182, 194
Spanish-American War, 194
Spiegel, George, 36, 37
Sporn, Philip, 41, 42
Standard Oil of New Jersey, 43
Steam, natural, 240-41
Steinfeld, Jesse L., 94
Sternglass, Ernest J., 22, 23, 48, 49, 59-60, 63-83, 85, 97, 118-21, 137, 213, 268
Stevens, John F., 178
Stewart, Alice, 62, 63, 67, 76, 87, 90, 92, 117-18, 121
Storage, of radioactive wastes, 148-171
Storer, John, 71, 72, 82
Strontium-90. See Radiation
Suez Canal crisis (1956), 174
Suffolk County Scientists for Cleaner Power and a Safer Environment, 116, 118-19
Sulky test, 187

Tamplin, Arthur R., 22, 23, 48-49, 50, 79-80, 81-93, 96-104, 144, 255-57
Taylor, Lauriston S., 22-23, 53, 54, 89, 92-93
Teller, Edward, 96, 175, 228-29
Tennessee-Tombigbee Water Project, 176
Tennessee Valley Authority (TVA), 258, 262

Tewksbury, Betty, 10
Thompson, Theos J., 286
Tlatelaco, Treaty of, 192
Tompkins, Paul C., 51
Totter, John R., 81-83
Tsivoglou, Ernest C., 126-27, 128, 132, 134
Tucker, Charles R. (Sonny), 273

Underground nuclear blasting. *See* Project Plowshare
Union of Concerned Scientists, 17
United Kingdom. *See* Great Britain
United Nations Conference on the Human Environment, 295
United States, 233, 235, 242; and atmospheric testing, 63-66; and Panama, 193-96
U.S. Army Corps of Engineers, 164-165, 185-86
U.S. Army Surgeon-General, 190
U.S. Congress, 30, 46, 126, 178, 179, 192, 222, 230, 296
U.S. Department of Defense, 24, 115
U.S. Department of Health, Education and Welfare, 69, 94
U.S. House of Representatives, 167-169
U.S. Interior Department, 141, 166
U.S. Justice Department, 39

U.S. Public Health Service, 6-7, 70, 205, 222
U.S. Senate, 169-70
U.S. Supreme Court, 126
U.S.S.R. *See* Soviet Union

Vermont Yankee Nuclear Corporation, 31-34

Walters, Barbara, 72
Warren, Shields, 68
Water Quality Improvement Act (1970), 260, 264, 279, 295
Watkins, J. Wade, 202
Watson, James D., 116-17
Watt, James, 90
Weinberg, Alvin M., 152-53
Western Europe, 242, 290
Westinghouse, 40-41, 134
Whisky Rebellion of 1794, 217
White, Irvin L., 288
Wilkins, M. H. F., 116
Wright, James Skelly, 282-85

X-rays, and leukemia, 61-64, 87, 117, 121; *see also* Radiation

Yoder, Franklin, 74, 75
Yook Ng, 102
Youngdahl, Russell C., 137